D1315317

Engineering Reliability

NEW TECHNIQUES AND APPLICATIONS

B. S. Dhillon
Department of Mechanical Engineering
University of Ottawa

Chanan Singh
Department of Electrical Engineering
Texas A & M University

A WILEY-INTERSCIENCE PUBLICATION

JOHN WILEY & SONS, New York · Chicester · Brisbane · Toronto

Library of Congress Cataloging in Publication Data
Dhillon, B S
   Engineering reliability.

   (Wiley series on systems engineering and analysis.)
   "A Wiley-Interscience publication."
   Includes bibliographies and index.
   1. Reliability (Engineering)   I. Singh, Chanan,
joint author. II. Title.
TS173.D48   620:'.00452   80-18734
ISBN 0-471-05014-8

Printed in the United States of America
10 9 8 7 6 5 4 3 2 1

SYSTEMS ENGINEERING AND ANALYSIS SERIES

In a society which is producing more people, more materials, more things, and more information than ever before, systems engineering is indispensable in meeting the challenge of complexity. This series of books is an attempt to bring together in a complementary as well as unified fashion the many specialties of the subject, such as modeling and simulation, computing, control, probability and statistics, optimization, reliability, and economics, and to emphasize the interrelationship among them.

The aim is to make the series as comprehensive as possible without dwelling on the myriad details of each specialty and at the same time to provide a broad basic framework on which to build these details. The design of these books will be fundamental in nature to meet the needs of students and engineers and to insure they remain of lasting interest and importance.

Preface

The scope of reliability engineering is extremely wide, encompassing many areas of engineering technology. Reliability engineering helps ensure the success of space missions, maintain the national security, deliver a steady supply of electric power, provide reliable transportation, and so on. There has been a considerable growth of knowledge in several areas of reliability engineering and its applications. These areas are characterized either by specific methodologies—fault trees, for example—that have found applications across various disciplines, or by topics that have developed a structure of their own, like power system reliability. New definitions, concepts, and techniques have been developed in these areas, and the knowledge of generic reliability theory alone is not enough for the appreciation of these ideas.

Reliability engineers deal with projects relating to various disciplines or with discrete aspects of a complex project and need the knowledge of diverse topics. An engineer needing information in these areas generally faces a great deal of difficulty and inconvenience, since these topics are discussed in various technical papers and in specialized books but have not been treated within the framework of a single book. This book is intended to fulfill the need for a single volume that considers these diverse topics. In this book topics of current interest are treated in such a manner that the reader needs no previous knowledge to understand the contents. We have tried to focus more on the structure of the concepts than the minute details. References to relevent literature are provided for the reader who wants to delve more deeply into particular topics.

The first chapter reviews the role and importance of reliability engineering in the planning and design process and outlines the scope of the book. Chapter 2 reviews the basic probability theory and other pertinent mathematical topics. Fundamental concepts and reliability techniques are described in Chapter 3. For readers not familiar with the basic concepts of reliability theory, these two chapters provide sufficient background for understanding this book.

Subsequent chapters deal with important techniques and specific areas of application. These chapters are self-contained and the reader with some background in reliability can understand them without referring to Chapters 2 and 3. Readers new to this area should find Chapters 2 and 3 helpful.

Chapter 4 presents the important and useful techniques of fault-tree analysis and common-cause failures. These two topics have been of considerable interest in recent years. Software reliability is discussed in Chapter 5, which describes the models and techniques for assessing and enhancing software reliability. The commonly used models and techniques for studies of mechanical and human reliability are presented in Chapters 6 and 7, respectively. Chapter 8 contains the reliability evaluation techniques and models for networks comprised of devices with two mutually exclusive failure modes. Markov models of repairable components are also described in this chapter.

Chapters 9 to 11 present three significant areas of application, electric power systems, transit systems, and computer systems. These areas of application have attracted a considerable amount of attention and have seen a substantial growth of knowledge. The reader will find a certain commonality of concepts but a great diversity in definitions, models, and methods.

The book is intended primarily for engineers, managers, graduate students, and other professionals interested in the subject of reliability. It can be adopted for a variety of graduate or short professional courses. A general course in reliability engineering would focus on Chapters 2 to 4, 7, 8, and selected portions of the remaining chapters. A course in power systems reliability could be based on Chapters 2, 3, and 9. Chapters 2, 3, and 10 could be used for a course on the reliability of urban transportations systems; and a computer systems reliability course would use Chapters 2, 3, 5, and 11.

Our experience on many projects and environments, teaching, and exposure to several outstanding experts in this area filter through the pages of this book. We would specifically like to thank our former colleagues and fellow professionals at Ontario Hydro and the Ontario Ministry of Transportation and Communications, and our present colleagues at the University of Ottawa and Texas A & M University, as well as many other professionals who, through discussion and writing, have influenced our thinking. We would also like to thank the Department of Electrical Engineering, Texas A & M University, for assistance during the preparation of the manuscript.

We thank our wives, Rosy Dhillon and Gurdeep Singh, for their patience and ever present help during the preparation of the manuscript, and we appreciate the support and encouragement of our parents throughout.

B. S. DHILLON
C. SINGH

Ottawa, Ontario
College Station, Texas
August 1980

Contents

Engineering Reliability

1

Introduction

1.1 RELIABILITY ASSURANCE

Reliability is an important consideration in the planning, design, and operation of systems. People have always expected trains to be on time, electric power not to fail, and so on. Before the Second World War, the concept of reliability had been only intuitive, subjective, and qualitative. The concept of quantitative reliability appears to have had its inception during the Second World War, and continues today, required by the size and complexity of modern systems.

The modern discipline of reliability is distinguished from the old concept by quantitative evaluation versus the older qualitative evaluation. When reliability is defined quantitatively, it is specified, analyzed, and measured and becomes a parameter of design that can be traded off against other parameters such as cost and performance [1].

The modern discipline of reliability had its origins in the military and space technology. Its influence has been steadily spreading into many other applications. This again is due to the growing complexity of systems, competitiveness in the market, and an ever-increasing competition for budget and resources. Neither can unreliability be tolerated nor are over designs permissible in todays market. The cost of failures in modern power systems and urban transportation system goes much beyond the cost of repair or replacement of effected parts. The inconvenience to consumers and commuters, lost products, crime, and decreased productivity cost much more than the price of immediate repairs.

1.2 PLANNING AND DESIGN

Quantitative reliability can play an important role in the planning, design, and operation of any system. As an example, consider transit facilities being planned for a city. The reliability characteristics of the vehicles and other equipment should be considered at an early stage [2]. The number of vehicles on scheduled maintenance and the number of vehicles that require service by failures should be allowed for while estimating the fleet size. A quantitative and consistent approach would be first to decide the level of service reliability with which the demand should be met and then develop

1

a system reliability model utilizing failure rate data [2] to estimate the number of spare vehicles.

Another approach considers reliability both as a constituent cost and as an effectiveness constraint in assessing the total cost of system acquisition and ownership, that is, life cycle costing. There is a financial penalty associated with vehicle failures since they must be repaired and other vehicles must be available in reserve to maintain the required service level. Also if the system reliability is not adequate, it can lead to loss of revenue because of reduced ridership. On the other hand, it generally costs more money to build higher reliability into the system. Therefore, a trade off can be made between cost and reliability.

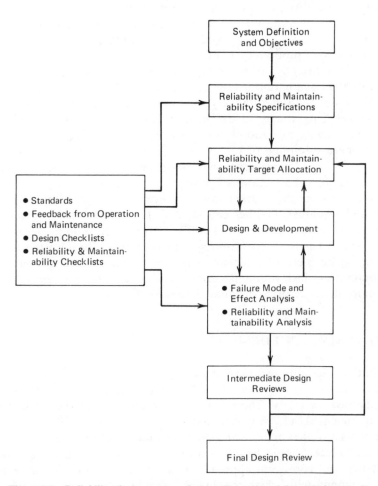

Figure 1.1   Reliability during system design. (Reproduced from Reference 2)

The inherent level of reliability is built into a system during its design phase [3]. Lack of control and direction during this period can result in costly retrofits or poor service reliability during the life cycle of the system [2]. The role of reliability engineering during the design process [2] is indicated in Figure 1.1.

1.3 TECHNIQUES AND APPLICATIONS

The body of knowledge regarding the theory and practice of reliability has been steadily growing. Not only has the basic reliability theory [3] become more sophisticated, but relatively new techniques have been developed and the areas of application considerably expanded. Techniques like fault trees have found applications across various disciplines. The topic of three-state devices is of great interest to reliability engineers in various disciplines. Software reliability, mechanical reliability, and human reliability may have borrowed some concepts from traditional hardware reliability but have distinguishing characteristics and concepts of their own. The concept of an error or a bug, for example, is different from the hardware failures and mechanical reliability is uniquely based on interference models. The areas of application like computers, power systems, and transit systems have developed their own definitions, concepts, and techniques. There is a certain commonality of concepts but great diversity in definitions, models, and methods. The forced outage rate, for example, when applied to generating units means the unavailability of the unit. The degree of growth of knowledge in particular areas can be seen by the fact that there are at present three books in English alone on power system reliability.

1.4 SCOPE

Engineers today dealing with large and diverse projects require information on reliability as it affects differing systems. These topics are discussed either in various technical papers or in specialized books and are presently not treated within the framework of a single book. An engineer needing information in these areas generally faces a great deal of difficulty. This book is an attempt to fulfill this need by treating these diverse topics in a single volume. Previous knowledge is not necessary to understand the contents, since two chapters on basic reliability theory are provided to give enough background. This book will find application in many disciplines and will be especially useful to reliability engineers, system engineers, and students of reliability.

REFERENCES

1. Singh, C., and M. D. Kankam, "Failure Data Analysis for Transit Vehicles," *Proceedings of the 1979 Annual Reliability and Maintainability Symposium*, IEEE, New York, 1979.

2. Singh, C., and M. D. Kankam, "Reliability Data and Analysis for Transit Vehicles," Research Report No. RR217, Research and Development Division, Ministry of Transportation and Communications, Ontario, Canada, Jan. 1977.

3. Singh, C., and R. Billinton, *System Reliability Modelling and Evaluation*, Hutchinson Educational, London, 1977.

2

Reliability Mathematics

2.1 INTRODUCTION

This chapter presents some basic mathematical concepts needed to understand subsequent chapters. Such topics as set theory, discrete and continuous random variables, probability distributions, hazard plotting, and differential equations are discussed briefly and provide an overview of the subject. The reader requiring in-depth knowledge of these concepts should consult references 1 and 5–7.

2.2 SET THEORY

Sets are normally represented by capital letters such as X, Y, Z. Elements are denoted by the lower case letters such as c, d, e.

If k is an element of set B, then it is denoted as: $k \in B$ and its negation is denoted as $k \notin B$. If X is a subset of set Y it is written as

$$X \subset Y \quad \text{or} \quad Y \supset X \tag{2.1}$$

The negation of the above is written as

$$X \not\subset Y \quad \text{or} \quad Y \not\supset X \tag{2.2}$$

If two sets are equal (suppose each set belongs to the other) they are expressed as

$$X = Y \tag{2.3}$$

The statement (2.3) is true if only

$$X \subset Y \quad \text{and} \quad Y \subset X \tag{2.4}$$

2.2.1 Union of Sets

The union of sets is denoted by the symbol $\cup$ or $+$. For example if $X + Y = Z$, it means that all the elements in set X or in set Y or in both sets

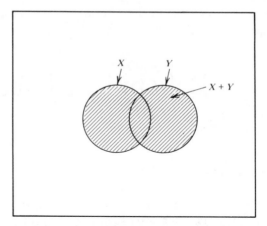

Figure 2.1 Venn diagram for the union of sets X, Y.

X and Y are contained in set Z. The statement

$$Z = X + Y \tag{2.5}$$

may also be written as $Z = X \cup Y$.

This case may be represented on the Venn diagram as shown in Figure 2.1.

2.2.2 Intersection of Sets

The intersection of sets is denoted by $\cap$ or dot ($\cdot$). For example, if the intersection of sets or events C and D is represented by a third set, say T, then this set contains all elements which belong to both C and D. It is

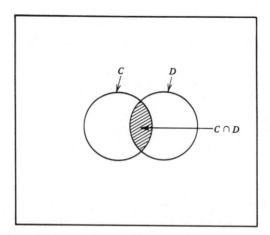

Figure 2.2 Venn diagram for intersection.

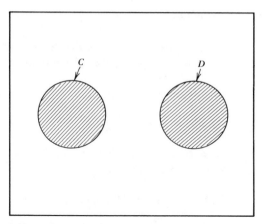

Figure 2.3 Venn diagram for disjoint sets C and D.

denoted as

$$T = C \cap D \quad \text{or} \quad T = C \cdot D \qquad (2.6)$$

The above expression is shown on the Venn diagram in Figure 2.2.

If the intersection of sets C and D is zero then sets C and D are called mutually exclusive or disjoint sets. This may be represented on Venn diagram as shown in Figure 2.3.

2.2.3 Basic Laws of Boolean Algebra

Some laws of Boolean algebra are as follows:

1. Distributive laws

$$X(Y+Z) = (X \cdot Y) + (X \cdot Z) \qquad (2.7)$$

$$X + (Y \cdot Z) = (X+Y) \cdot (X+Z) \qquad (2.8)$$

2. Boolean identities

$$X + X = X \qquad (2.9)$$

$$X \cdot X = X \qquad (2.10)$$

3. Absorption laws

$$X + (X \cdot Y) = X \qquad (2.11)$$

$$X(X \cdot Y) = X \cdot Y \qquad (2.12)$$

2.3 PROBABILITY THEORY

Probability theory may be defined as the study of the random experiments. The most important event-related properties of probability are as follows:

1. For each event X, the event probability is

$$0 \leqslant P(X) \leqslant 1 \tag{2.13}$$

2. In the case of mutually exclusive events, say $x_1, x_2, x_3, \ldots, x_n$, the probability of union of events is given by

$$P(x_1 + x_2 + x_3 + \ldots + x_n) = P(x_1) + P(x_2)$$
$$+ P(x_3) + \ldots + P(x_n) \tag{2.14}$$

3. The union of n events is given by

$$P(x_1 + x_2 + x_3 + \ldots + x_n) = \{P(x_1) + P(x_2) + \ldots + P(x_n)\}$$
$$- \{P(x_1 x_2) + P(x_1 x_3) + \ldots + P(x_j x_{ij \neq i})\} + \ldots$$
$$+ (-1)^{n-1} \{P(x_1 x_2 x_3 \ldots x_n)\} \tag{2.15}$$

For example, in the case of two statistically independent events x_1 and x_2, the probability expressions becomes:

$$P(x_1 + x_2) = P(x_1) + P(x_2) - P(x_1)P(x_2) \tag{2.16}$$

4. Probability of the sample space S is always equal to unity, that is,

$$P(S) = 1 \tag{2.17}$$

The negation of the sample space S is written as $\bar{S}$. Thus

$$P(\bar{S}) = 0 \tag{2.18}$$

5. The n events intersection probability expression is as follows

$$P(x_1 x_2 x_3 \ldots x_n) = P(x_1)P(x_2/x_1) \ldots P(x_n/x_1 x_2 \ldots x_{n-1}) \tag{2.19}$$

where $P(x_2/x_1)$ implies probability of x_2 given x_1.

If all the events are statistically independent, the above expression becomes

$$P(x_1 x_2 x_3 \ldots x_n) = P(x_1)P(x_2)P(x_3) \ldots P(x_n) \tag{2.20}$$

6. The events X and Y are said to be independent, if and only if

$$P(XY) = P(X)P(Y) \tag{2.21}$$

If events X and Y cannot satisfy the above relationship, then these events are said to be dependent. The conditional probability of x_n, given that the events $x_1, x_2, x_3, \ldots, x_{n-1}$ have occurred is obtained by the following relationship:

$$P(x_n/x_1, x_2, x_3, \ldots, x_{n-1}) = \frac{P(x_1, x_2, x_3, \ldots, x_n)}{P(x_1, x_2, x_3, \ldots, x_{n-1})} \tag{2.22}$$

2.4 RANDOM VARIABLES

Random variables may be discrete or continuous. Both discrete and continuous variables and the associated probability distributions are described in these sections.

2.4.1 *Discrete Random Variables*

If Y is a random variable on the sample space S along with a countably infinite set $Y(S) = \{y_1, y_2, y_2, \ldots\}$, then these random variables along with other finite sets are known as discrete random variables.

Density Function.	For a single-dimension discrete random variable Y, the discrete probability function of the random variable Y is represented by $f(y_i)$ if the following conditions hold:

$$f(y_i) \geqslant 0 \qquad \text{for all} \qquad y_i \in R_y \text{ (range space)} \tag{2.23}$$

and

$$\sum_{\substack{\text{all} \\ y_i}} f(y_i) = 1 \tag{2.24}$$

Cumulative Probability Distribution Function.	The cumulative probability distribution function is defined as

$$F(y) = \sum_{y_i < y} f(y_i) \tag{2.25}$$

where $F(y)$ is the cumulative probability distribution function.

Furthermore, the area under the probability density function curve is always

$$0 \leqslant F(y) \leqslant 1 \tag{2.26}$$

Binomial Distribution. The binomial distribution is a frequently used distribution in reliability engineering. This is also known as the Bernoulli distribution. We are often concerned with the probabilities of outcome such as the total number of failures in a sequence of n trials. For this distribution, each trial has two possible outcomes, success and failure, where the probability of each trial remains constant.

The binomial probability function $f(x)$ is defined as

$$f(x) = \frac{n!}{x!(n-x)!} p^x q^{n-x}, \qquad x = 0, 1, 2, \ldots, n \qquad (2.27)$$

where x = the number of failures in n trials
  p = the single trial probability of success
  q = the single trial probability of failure

It is always true that the summation of probability of failure and success for each trial is always equal to unity (i.e., $p + q = 1$).

The probability of x or less failures in n number of trials is known as the probability distribution function, $F(x)$, $i \cdot e$.

$$F(x) = \sum_{i=0}^{x} \binom{n}{i} p^i q^{n-i} \qquad (2.28)$$

where $\binom{n}{i} = n!/i!(n-i)!$.

Poisson Distribution. This distribution model is used in reliability studies when one is interested in the occurrence of a number of events that are of the same kind. Occurrence of each event is represented as a point on a time scale. In reliability engineering each event represents a failure. The Poisson density function is defined as

$$f(n) = \frac{(\lambda t)^n \exp(-\lambda t)}{n!}, \qquad n = 0, 1, 2, \ldots \qquad (2.29)$$

where t is the time and λ is the constant failure or arrival rate.

The cumulative distribution function F is given by

$$F = \sum_{i=0}^{n} \frac{(\lambda t)^i \exp(-\lambda t)}{i!} \qquad (2.30)$$

Multinomial Distribution. This distribution is applicable to those cases where a system or device has more than two states. This is an extension of the binomial distribution which is only applicable to systems or devices with two states. The multinomial distribution probability function is defined as follows:

$$f(x_1, x_2, x_3, \ldots, x_n) = \frac{n!}{x_1! x_2! x_3! \ldots x_n!} P_1^{x_1} P_2^{x_2} P_3^{x_3} \ldots P_n^{x_n} \quad (2.31)$$

for
$$\left. \begin{array}{c} \sum_{i=1}^{n} P_i = 1 \\ \sum_{i=1}^{n} x_i = n \end{array} \right\} \quad 0 < P_i < 1$$

2.4.2 Continuous Random Variables

A real-valued function defined over a sample space S is called a continuous random variable. In the case of the continuous random variable, the probability density function is defined as

$$f(t) = \frac{dF(t)}{dt} \quad (2.32)$$

where

$$F(t) = \int_{-\infty}^{t} f(x)\, dx \quad (2.33)$$

and

$$F(\infty) = 1$$

$F(t)$ is called the distribution function of a continuous random variable t. The probability distributions of the continuous random variable are as follows:

Uniform Distribution. This is a continuous distribution whose probability density $f(t)$ and distribution functions $F(t)$, respectively, are defined as follows:

$$f(t) = \frac{1}{\alpha - \theta} \quad \theta < t < \alpha \quad (2.34)$$

otherwise

$$f(t) = 0$$

and

$$F(t) = \begin{cases} 1 & t \geqslant \alpha \\ 0 & t \leqslant \theta \\ \dfrac{t-\theta}{\alpha-\theta} & \theta < t < \alpha \end{cases} \qquad (2.35)$$

Exponential Distribution. This is a widely used distribution in reliability engineering [2]. It is one of the simplest distributions to perform reliability analysis. The exponential probability density function $f(t)$ is defined as follows:

$$f(t) = \lambda e^{-\lambda t} \qquad t \geqslant 0 \qquad \lambda > 0 \qquad (2.36)$$

where λ is a constant failure rate and t is time.

The cumulative distribution function $F(t)$ is given by

$$F(t) = 1 - e^{-\lambda t} \qquad (2.37)$$

Weibull Distribution. This distribution is due to Weibull [8]. This distribution can represent many different physical phenomena. Weibull distribution is a three parameters distribution whose probability density function is defined as follows:

$$f(t) = \frac{b}{n}(t-\alpha)^{b-1} e^{-\{(t-\alpha)^b/n\}} \qquad \text{for} \quad t > \alpha \qquad b, n, \alpha > 0 \quad (2.38)$$

where b, n, and α are shape, scale, and location parameters, respectively.

The distribution function is given by

$$F(t) = 1 - e^{-\{(t-\alpha)^b/n\}} \qquad \text{for} \quad t > \alpha \qquad n, b > 0 \qquad \alpha \geqslant 0 \quad (2.39)$$

Rayleigh Distribution. This distribution has its applications in the theory of sound and reliability engineering. The Rayleigh distribution is a special case of the Weibull distribution ($b = 2$, $\alpha = 0$). Therefore, the probability density and distribution functions may be directly obtained from (2.38) and (2.39), respectively, as follows:

$$f(t) = \frac{2}{n} t e^{-t^2/n} \qquad t \geqslant 0 \qquad n > 0 \qquad (2.40)$$

and

$$F(t) = 1 - e^{-t^2/n} \qquad (2.41)$$

Gamma Distribution. This distribution is an extension of the exponential distribution. Some of its applications are found in life test problems.

Probability density and distribution functions are

$$f(t) = \frac{\lambda(\lambda t)^{\alpha-1}}{\Gamma(\alpha)} e^{-\lambda t} \qquad t \geqslant 0 \qquad \lambda, \alpha > 0 \tag{2.42}$$

and

$$F(t) = 1 - \sum_{i=0}^{\alpha-1} \frac{e^{-\lambda t}(\lambda t)^i}{i!} \qquad t \geqslant 0 \qquad \lambda, \alpha > 0 \tag{2.43}$$

In the case of $\alpha = 1$, this distribution reduces to exponential form.

Extreme Value Distribution. It is a good representative of the failure behavior of mechanical components. Probability density and distribution function of the extreme value distribution are as follows:

$$f(t) = e^t e^{-e^t} \qquad -\infty < t < \infty \tag{2.44}$$

and

$$F(t) = 1 - e^{-e^t} \qquad -\infty < t < \infty \tag{2.45}$$

Normal Distribution (Gaussian). This is a two-parameter distribution, which also has its applications in the reliability field. Its probability density function is defined as follows:

$$f(t) = \frac{1}{\sqrt{2\pi}\,\sigma} e^{-1/2 \frac{(t-\mu)^2}{\sigma}} \qquad -\infty < t < \infty \qquad \sigma > 0 \qquad -\infty < \mu < \infty \tag{2.46}$$

The cumulative distribution function is

$$F(t) = \frac{1}{\sqrt{2\pi}\,\sigma} \int_{-\infty}^{t} e^{-1/2 \left(\frac{x-\mu}{\sigma}\right)^2} dt \tag{2.47}$$

The numerical values of the cumulative function (2.47) may be obtained from the standard tables.

Log Normal Distribution. This is another distribution often used to represent the repair times of failed equipment. The probability density and distribution functions are

$$f(t) = \frac{1}{(t-\alpha)\sqrt{2\pi}\,\sigma} e^{-\{\log_e(t-\alpha)-\mu\}^2/2\sigma^2} \qquad \text{for} \quad t > \alpha > 0 \qquad \sigma > 0 \tag{2.48}$$

and

$$F(t) = \frac{1}{\sqrt{2\pi}\,\sigma} \int_0^t \frac{1}{x} e^{-\{\ln x - \mu\}^2/2\sigma^2}\,dx \qquad \text{for} \quad t>0 \qquad (2.49)$$

Beta Distribution. The beta distribution is a two-parameter distribution finding some uses in reliability engineering. The probability density function of this distribution is defined as follows:

$$f(t) = \frac{(\gamma+\beta+1)!}{\gamma!\beta!} t^\gamma (1-t)^\beta$$

$$\text{for} \quad 0<t<1 \qquad \gamma>-1 \qquad \beta>-1 \qquad (2.50)$$

The cumulative distribution function is given by

$$F(t) = \int_0^t \frac{(\gamma+\beta+1)}{\gamma!\beta!} y^\gamma (1-y)^\beta\,dy \qquad \text{for} \quad 0<t<1 \qquad (2.51)$$

The General Distribution (Hazard-Rate Model). This section presents a general distribution [3] which might be useful to represent failure behavior of items that are not adequately represented by the existing failure distributions.

The hazard rate $\lambda(t)$ and reliability function $R(t)$ are defined by

$$\lambda(t) = k\lambda c t^{c-1} + (1-k)bt^{b-1}\beta e^{\beta t^b}$$

$$\text{for} \quad b,c,\beta,\lambda>0 \qquad 0\leqslant k\leqslant 1 \qquad t\geqslant 0 \qquad (2.52)$$

and

$$R(t) = \exp\left[-k\lambda t^c - (1-k)(e^{\beta t^b}-1)\right] \qquad (2.53)$$

where b,c = shape parameters
$\quad\quad\;\; \beta,\lambda$ = scale parameters
$\quad\quad\;\;\; t$ = time

In special cases, the above distribution becomes

$c=1,\, b=1$	Makeham distribution
$k=0,\, b=1$	extreme value
$k=1$	Weibull
$c=0.5,\, b=1$	bathtub curve

The Hazard Rate Model Distribution. The hazard rate function $\lambda(t)$ [4] of this model is defined as follows:

$$\lambda(t) = k\lambda \tanh \lambda t + (1-k)bt^{b-1}\beta e^{-\beta t^b}$$

$$\text{for} \quad b, \beta, \lambda > 0 \quad\quad 0 \leqslant k \leqslant 1 \quad\quad t \geqslant 0 \quad\quad\quad (2.54)$$

where b = the shape parameter
$\quad\quad\quad \beta, \lambda$ = the scale parameters

The reliability function is given by

$$R(t) = \exp\left\{ -k\ln\cosh \lambda t + (1-k)(e^{-\beta t^b} - 1)\right\} \quad\quad (2.55)$$

Figure 2.4 shows some selective curves ($\beta = \lambda = 1$) for the hazard rate function expressed in (2.54).

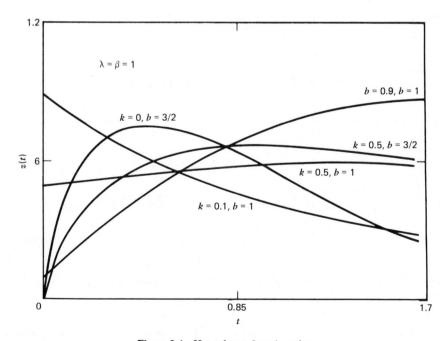

Figure 2.4 Hazard rate function plot.

2.5 EXPECTED VALUE AND VARIANCE OF THE RANDOM VARIABLES

The expected value, $E(x)$, of a continuous random variable is defined as

$$E(x) = \int_{-\infty}^{\infty} xf(x)\, dx \tag{2.56}$$

Similarly in the case of a discrete random variable x, the expected value, $E(x)$, is given by

$$E(x) = \sum_{i=1}^{k} x_i f(x_i) \tag{2.57}$$

where k are the discrete values of the random variable x.

The variance $\sigma^2(x)$ of a random variable x is defined by

$$\sigma^2(x) = E(x^2) - \{E(x)\}^2 \tag{2.58}$$

2.6 MOMENT-GENERATING FUNCTION

The moment-generating function, $M_t(\theta)$ is defined for both continuous and discrete cases as follows:

Continuous case

$$M_t(\theta) = \int_{-\infty}^{+\infty} \exp(\theta t) f(t)\, dt \tag{2.59}$$

and discrete case

$$M_t(\theta) = \sum_{k=1}^{n} \exp(\theta t_k) f(t_k)\, dt \tag{2.60}$$

for obtaining the nth moment about origin we apply the following:

$$\left[\frac{d^n M_t(\theta)}{d\theta^n} \right] \theta = 0 \tag{2.61}$$

Therefore, the expected value and variance are given by these relationships:

$$E(t) = \frac{dM_t(\theta)}{d\theta} \bigg|_{\theta=0} \tag{2.62}$$

and

$$E(t^2) = \frac{d^2 M_t(\theta)}{d\theta^2}\bigg|_{\theta=0} \tag{2.63}$$

2.7 HAZARD PLOTTING FOR INCOMPLETE FAILURE DATA

This is a graphical data analysis technique [7] to establish failure distributions for units with incomplete failure data. Failure data are complete failure data if the failure times for all units in a sample are contained. In contrast, the failure data are called incomplete failure data if a sample contains both the failure times of failed units and running times of unfailed units. The unfailed units running times are called censoring times.

In addition, if in a sample all the unfailed units under observation have different censoring times, then the failure data are called multiply censored. Furthermore, if the unfailed units in a sample have the same censoring time and in addition the censoring time is greater than the failure times, then the failure data are called singly censored. This type of data results when a sample of items undergo life testing and termination of testing before all units fail, whereas the multiply censored data result from any of the following:

1. From the operating units.
2. Some extraneous causes.
3. Units removal before failure.

Some of the advantages of this hazard plotting technique are as follows:

1. It provides a visibility tool because the pictorial plots are easy to grasp.
2. Data plots are an easy way to fit a theoretical distribution to data.
3. It simplifies for the analyst to assess the adequate fit of a theoretical distribution to data.

2.7.1 Hazard Rate Plotting Theory

This technique is based on the distribution hazard rate function concept. The following three basic relationships associated with the hazard plotting technique are defined as

$$z(t) = \frac{f(t)}{R(t)} = \frac{f(t)}{1 - F(t)} \tag{2.64}$$

where $z(t)$ = the hazard rate function
        $R(t)$ = the reliability function
        $F(t)$ = the failure distribution function

The cumulative hazard, $z_c(t)$, is given by

$$z_c(t) = \int_{-\infty}^{t} z(t)\,dt = -[\ln\{1 - F(t)\}] \tag{2.65}$$

The cumulative distribution function $F(t)$ is defined as

$$F(t) = 1 - e^{-z_c(t)} \tag{2.66}$$

The above relationship is very useful to determine hazard function properties.

2.7.2 Hazard Plotting for the Weibull Distribution

This example is presented for the Weibull distribution. However, interested readers should consult reference 7 for other distributions as well as for a detailed presentation of this approach. Here the theory behind the Weibull hazard plotting is briefly described.

The Weibull hazard, $z(t)$, and the density function, $f(t)$, are defined as follows:

$$f(t) = \frac{\alpha t^{\alpha-1}}{\beta^{\alpha}} e^{-(t/\beta)^{\alpha}} \qquad \alpha, \beta \geqslant 0 \qquad t \geqslant 0 \tag{2.67}$$

and

$$z(t) = \frac{\alpha t^{\alpha-1}}{\beta^{\alpha}} \tag{2.68}$$

Both cumulative distribution and hazard functions are obtained by integrating expressions (2.67) and (2.68) over the time interval $[0, t]$ as follows:

$$F(t) = 1 - e^{-(t/\beta)^{\alpha}} \tag{2.69}$$

and

$$z_c(t) = (t/\beta)^{\alpha} \tag{2.70}$$

By taking the $\log_e$ of (2.70) we get

$$\ln(t) = \alpha^{-1}\ln(z_c) + \ln(\beta) \tag{2.71}$$

The above equation indicated that the left-hand side of this expression is the linear function of $\ln(z_c)$, which indicates that the log-log graph paper is the Weibull hazard paper. Therefore, parameters α and β can be estimated graphically by using the log-log paper.

The shape parameter, α, is estimated from the fact that $1/\alpha$ is the slope of the straight line. At $z_c = 1$, the value of the β is equal to time t, therefore, by using this relationship, the value of the scale parameter β can be estimated.

2.8 LAPLACE TRANSFORMS

Some of these transforms are used in this book to solve systems of linear differential equations with constant coefficients. Furthermore, these transforms are applied in conjunction with other differential equation techniques to solve simpler type of partial differential equations. The basic definition of the Laplace transform $f(s)$, of a function $f(t)$ is as follows:

$$f(s) = \mathcal{L}\{f(t)\} = \int_0^\infty e^{-st} f(t) \, dt \tag{2.72}$$

where $s =$ the Laplace transform variable
$t =$ the time variable

Example 1.   Find the Laplace transform of the function $f(t) = t$, that is,

$$f(s) = \int_0^\infty e^{-st} t \, dt = \left[\frac{e^{-st}}{-s} \left(t + \frac{1}{s} \right) \right]_0^\infty$$

$$= \frac{1}{s^2} \qquad \text{for } s > 0 \tag{2.73}$$

Example 2.   If $f(t) = e^{at}$, the Laplace transform of this exponential function becomes

$$f(s) = \int_0^\infty e^{-st} e^{at} \, dt = \int_0^\infty e^{(a-s)t}$$

$$= \left[-\frac{1}{(s-a)} e^{-(s-a)t} \right]_0^\infty$$

$$= \frac{1}{s-a} \qquad \text{for } s > a \tag{2.74}$$

2.8.1   Laplace Theorem of Derivatives

If $\mathcal{L}\{f(t)\} = f(s)$, then

$$\mathcal{L}\left\{ \frac{df(t)}{dt} \right\} = sf(s) - f(0) \tag{2.75}$$

2.8.2   Laplace Transform Initial-Value Theorem

If the following limits exist, then the Abel's theorem is

$$\lim_{t \to 0} f(t) = \lim_{s \to \infty} sf(s) \qquad (2.76)$$

2.8.3   Laplace Transform Final-Value Theorem

Provided the following limits exist, then the final-value theorem may be stated as:

$$\lim_{t \to \infty} f(t) = \lim_{s \to 0} sf(s) \qquad (2.77)$$

Laplace Transform Table

$f(t)$	$f(s)$	
t	$1/s$	
$\dfrac{df(t)}{dt}$	$sf(s) - f(0)$	
$\dfrac{d^2 f(t)}{dt^2}$	$s^2 f(s) - sf(0) - f'(0)$	
1	$\dfrac{1}{s}$	$s > 0$
e^{at}	$\dfrac{1}{s-a}$	$s > a$
t^k	$\dfrac{k!}{s^{k+1}}$	$s > 0$

2.9   PARTIAL FRACTION TECHNIQUE

This is used when finding inverse Laplace transforms of a rational function such as $G(s)/Q(s)$, where $G(s)$ and $Q(s)$ are polynomials, and the degree of $G(s)$ is less than that of $Q(s)$. Therefore, the ratio of $G(s)/Q(s)$ may be written as the sum of rational functions or partial fractions in the following forms:

$$\frac{A}{(\alpha s + \beta)^n}, \ \frac{Bs + C}{(\alpha s^2 + \beta s + C)^n} \qquad n = 1, 2, 3, \ldots$$

Heaviside Theorem.   This is used to obtain partial fractions and inverse of a rational function, $G(s)/Q(s)$.

The inverse of $G(s)/Q(s)$ may be written as:

$$\mathcal{L}^{-1} \left\{ \frac{G(s)}{Q(s)} \right\} = \sum_{i=1}^{k} \frac{G(\beta_i)}{Q'(\beta_i)} e^{\beta_i t} \qquad (2.78)$$

where the prime represents derivative with respect to s, β_i represents ith zero and k denotes total number of distinct zeros of $Q(s)$.

Example 3. Suppose

$$\frac{G(s)}{Q(s)} = \frac{s+2}{(s-4)(s-6)}$$

find the inverse Laplace transform. Hence,

$$G(s) = s+2 \qquad Q(s) = s^2 - 10s + 24 \qquad Q'(s) = 2s - 10$$

$$\beta_1 = 4 \qquad \beta_2 = 6 \qquad k = 2$$

Therefore,

$$\frac{G(4)}{Q'(4)} e^{4t} + \frac{G(6)}{Q'(6)} e^{6t} = 4e^{6t} - 3e^{4t} \tag{2.79}$$

2.10 DIFFERENTIAL EQUATIONS

The single-independent-variable linear first-order differential equations in the reliability study are mainly associated with the Markov technique. In this section we discuss how to solve such equations using integration techniques.

The first-order first-degree linear differential equation may be written in the following form

$$\frac{dP}{dt} + PG(t) = Q(t) \tag{2.80}$$

Since

$$\frac{d}{dt}\left(Pe^{\int G(t)\,dt}\right) = \frac{dP}{dt} e^{\int G(t)\,dt} + PG(t)e^{\int G(t)\,dt}$$

$$= e^{\int G(t)\,dt}\left[\frac{dP}{dt} + PG(t)\right] \tag{2.81}$$

The above expression shows that $e^{\int G(t)\,dt}$ is an integrating factor of the differential equation (2.80).

The primitive of differential equation (2.80) may be written as

$$Pe^{\int G(t)\,dt} = \int Q(t)e^{\int G(t)\,dt}\,dt + c \tag{2.82}$$

where c is a constant.

Example 4.   Obtain the solution equation for the following differential equation:

$$\frac{dP}{dt} + 6P = 8 \tag{2.83}$$

Hence,

$$e^{\int G(t)\,dt} = e^{\int 6\,dt} = e^{6t} \tag{2.84}$$

By substituting (2.84) into (2.82), we get

$$Pe^{6t} = 8 \int e^{6t}\,dt + c$$

$$P = \tfrac{8}{6} + ce^{-6t} \tag{2.85}$$

For given initial conditions; at $t = 0$, $P = 1$; the following value for the constant c is obtained from (2.85).

$$c = -\tfrac{1}{3}$$

$$P = \tfrac{4}{3} - \tfrac{1}{3}e^{-6t} \tag{2.86}$$

2.10.1   Differential Equation

Solution with Laplace Transform Technique.   Solving the same differential equation

$$\frac{dP}{dt} + 6P = 8 \tag{2.87}$$

with the Laplace transform method for same initial conditions we get

$$sP - 1 + 6P = \frac{8}{s}$$

$$P = \frac{8}{s(8+6)} + \frac{1}{s+6} \tag{2.88}$$

The inverse Laplace transform of the above equation is

$$P = \tfrac{4}{3} - \tfrac{1}{3}e^{-6t} \tag{2.89}$$

This shows that solution (2.86) is same as the solution to (2.89).

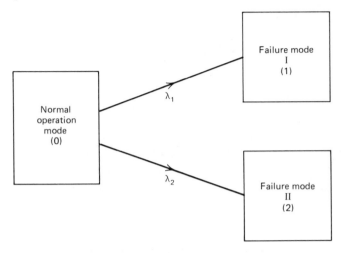

Figure 2.5   State space diagram.

2.10.2   System of Linear First Order Differential Equations

The following system of linear first-order differential equations with constant coefficients are associated with the transition diagram* of Figure 2.5. This transition diagram represents a three-state device, for example, a fluid flow valve, electronic diode, an electrical switch, etc.

$$\frac{dP_0(t)}{dt} + (\lambda_1 + \lambda_2)P_0(t) = 0 \tag{2.90}$$

$$\frac{dP_1(t)}{dt} - \lambda_1 P_0(t) = 0 \tag{2.91}$$

$$\frac{dP_2(t)}{dt} - \lambda_2 P_0(t) = 0 \tag{2.92}$$

At $t=0$, $P_0(t)=1$, and other probabilities are zero.

The Laplace transforms of differential equations (2.90)–(2.92) are

$$(s+\lambda_1+\lambda_2)P_0(s) \quad 0P_1(s) \quad 0P_2(s) = P_0(0) \tag{2.93}$$

$$-\lambda_1 P_0(s) \quad sP_1(s) \quad 0P_2(s) = P_1(0) \tag{2.94}$$

$$-\lambda_2 P_0(s) \quad 0P_1(s) \quad P_2(s) = P_2(0) \tag{2.95}$$

$$\begin{bmatrix} s+\lambda_1+\lambda_2 & 0 & 0 \\ -\lambda_1 & s & 0 \\ -\lambda_2 & 0 & s \end{bmatrix} \begin{bmatrix} P_0(s) \\ P_1(s) \\ P_2(s) \end{bmatrix} = \begin{bmatrix} 1 \\ 0 \\ 0 \end{bmatrix}$$

*Markov technique is discussed in Chapter 3.

By solving the above we get

$$P_0(s) = \frac{1}{s+\lambda_1+\lambda_2} \tag{2.96}$$

$$P_1(s) = \frac{\lambda_1}{s(s+\lambda_1+\lambda_2)} \tag{2.97}$$

$$P_2(s) = \frac{\lambda_2}{s(s+\lambda_1+\lambda_2)} \tag{2.98}$$

The inverse Laplace transforms of (2.96)–(2.98) are

$$P_0(t) = e^{-(\lambda_1+\lambda_2)t} \tag{2.99}$$

$$P_1(t) = \frac{\lambda_1}{\lambda_1+\lambda_2}(1 - e^{-(\lambda_1+\lambda_2)t}) \tag{2.100}$$

$$P_2(t) = \frac{\lambda_2}{\lambda_1+\lambda_2}(1 - e^{-(\lambda_1+\lambda_2)t}) \tag{2.101}$$

REFERENCES

1. Bean, H. S., *Differential Equations*, Addison-Wesley, Reading, MA, 1962.
2. Davis, D. J., "An Analysis of Some Failure Data," *J. Amer. Stat. Assoc.*, 113–150 (June 1952).
3. Dhillon, B. S., "A Hazard Rate Model," *IEEE Trans. Reliab.*, **R-29**, 150 pp. (1979).
4. Dhillon, B. S., "New Hazard Rate Functions," *Microelectron. Reliab.* (1978).
5. Drake, A., *Fundamental of Applied Probability Theory*, McGraw-Hill, New York, 1967.
6. Feller, W., *An Introduction to Probability Theory and Its Applications*, Vol. I, Wiley, New York, 1957.
7. Nelson, W., "Hazard Plotting for Incomplete Failure Data," *J. Qual. Tech.*, **1** (1) (Jan. 1969).
8. Weibull, W., "A Statistical Distribution Function of Wide Applicability," *J. Appl. Mech.*, **18**, 293–297 (March 1951).

3

Fundamental Concepts in Reliability Engineering

3.1 INTRODUCTION

This chapter briefly presents the fundamental concepts in reliability engineering such as general reliability function, redundant networks, reliability evaluation techniques, reliability apportionment, and failure mode and effect analysis.

In this chapter, a brief discussion on reliability evaluation techniques such as binomial, Markov processes (state space approach), decomposition, minimal cut set, and network reduction is presented. The delta-star technique is presented in a more detailed form.

3.2 GENERAL RELIABILITY FUNCTION

3.2.1 General Concepts

Suppose n_0 identical components are under test, after time t, $n_f(t)$ fail and $n_s(t)$ survive. The reliability function $R(t)$ is defined by

$$R(t) = \frac{n_s(t)}{n_s(t) + n_f(t)} \tag{3.1}$$

since

$$n_s(t) + n_f(t) = n_0$$

the equation becomes

$$R(t) = \frac{n_s(t)}{n_0} \tag{3.2}$$

since

$$R(t) + F(t) = 1$$

$$R(t) = 1 - F(t) \tag{3.3}$$

where $F(t)$ the failure probability at time t. To obtain failure probability $F(t)$, substitute (3.2) into (3.3), and subtract it from unity, that is,

$$F(t) = 1 - \frac{n_s(t)}{n_0}$$

since

$$n_s(t) + n_f(t) = n_0$$

$$F(t) = 1 - \left(\frac{n_0 - n_f(t)}{n_0} \right) = \frac{n_f(t)}{n_0} \tag{3.4}$$

By using the above result in relationship (3.3) we get

$$R(t) = 1 - F(t) = 1 - \frac{n_f(t)}{n_0} \tag{3.5}$$

the derivative of (3.5) with respect to time t is

$$\frac{dR(t)}{dt} = -\frac{1}{n_0} \frac{dn_f(t)}{dt} \tag{3.6}$$

In the limiting case, as dt approaches zero, the expression (3.6) is the instantaneous failure density function $f(t)$, that is,

$$\frac{1}{n_0} \frac{dn_f(t)}{dt} \to f(t)$$

Therefore, expression (3.6) becomes

$$\frac{dR(t)}{dt} = -f(t) \tag{3.7}$$

By using relationship (3.2) the other form of (3.6) may be written as

$$\frac{dn_f(t)}{dt} = -n_0 \frac{dR(t)}{dt} = \frac{dn_s(t)}{dt} \tag{3.8}$$

3.2.2 Instantaneous Failure or Hazard Rate

If we divide both sides of (3.8) by $n_s(t)$, we get

$$\frac{1}{n_s(t)}\frac{dn_f(t)}{dt} = \frac{-n_0}{n_s(t)}\frac{dR(t)}{dt} \tag{3.9}$$

This is equal to the hazard rate $\lambda(t)$, that is,

$$\frac{1}{n_s(t)}\frac{dn_f(t)}{dt} = -\frac{n_0}{n_s(t)}\frac{dR(t)}{dt} = \lambda(t) \tag{3.10}$$

By substituting (3.2) and (3.7) into (3.10), we get an expression for the hazard rate (instantaneous rate):

$$\lambda(t) = -\frac{1}{R(t)}\frac{dR(t)}{dt} = \frac{f(t)}{R(t)} \tag{3.11}$$

3.2.3 Reliability Function

Equation 3.11 may be rewritten in the following form:

$$\frac{-dR(t)}{R(t)} = \lambda(t)\,dt \tag{3.12}$$

By integrating both sides of (3.12) over the time range 0 to t, we get

$$\int_0^t \lambda(t)\,dt = -\int_1^{R(t)} \frac{1}{R(t)}\,dR(t)$$

For the known initial condition that at $t=0$, $R(t)=1$ the above integral expression becomes

$$\ln R(t) = -\int_0^t \lambda(t)\,dt \tag{3.13}$$

The following general reliability function is obtained from (3.13):

$$R(t) = e^{-\int_0^t \lambda(t)\,dt} \tag{3.14}$$

Where $\lambda(t)$ is the time-dependent failure rate or instantaneous failure rate. It is also called the hazard rate. The above expression is a general reliability function. In other words, it can be used to obtain a component reliability for any known failure time distribution.

3.3   BATHTUB HAZARD RATE CURVE

This hazard rate curve, shown in Figure 3.1, is regarded as a typical hazard rate curve, especially when representing the failure behavior of electronic components. Mechanical components may or may not follow this type of failure pattern.

As shown in Figure 3.1 the decreasing hazard rate is sometimes called the "burn-in period." There are also several other names for this period such as debugging period, infant mortality period, break-in period. Occurrence of failures during this period is normally attributed to design or manufacturing defects.

The constant part of this bathtub hazard rate is called the "useful period," which begins just after the infant mortality period and ends just before the "wear-out period."

The wear-out period begins when an equipment or component has aged or bypassed its useful operating life. Consequently, the number of failures during this time begin to increase. Failures that occur during the useful life are called "random failures" because they occur randomly or in another word unpredictably.

The hazard rate shown in Figure 3.1 can be represented by the following function [15]:

$$\lambda(t) = k\lambda c t^{c-1} + (1-k)bt^{b-1}\beta e^{\beta t^b} \tag{3.15}$$

for $b, c, \beta, \lambda > 0$ $0 \leqslant k \leqslant 1$ $t \geqslant 0$ and $c = 0.5$ $b = 1$

where $b, c =$ shape parameters
        $\beta, \lambda =$ scale parameters
        $t =$ time

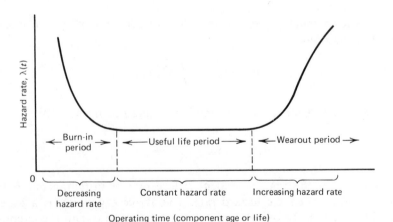

Figure 3.1   Bathtub curve.

3.4 MEAN-TIME-TO-FAILURE (MTTF)

The expected value $E(t)$, in our case MTTF, of a probability density function of the continuous random variable time t is given by

$$E(t) = \text{MTTF} = \int_0^\infty tf(t)\, dt \qquad (3.16)$$

where $f(t)$ is the failure density function.

Example 1. Suppose, a component failure time follows the exponential failure law. It follows that the component has constant failure rate, λ (i.e., useful life period of the bathtub curve). Find the reliability function and the mean-time-to-failure expressions.

From the known information,

$$f(t) = \lambda e^{-\lambda t} \qquad (3.17)$$

and

$$\lambda(t) = \lambda \qquad (3.18)$$

To obtain the reliability function substitute (3.18) into (3.19):

$$R(t) = e^{-\int_0^t \lambda\, dt} = e^{-\lambda t} \qquad (3.19)$$

In the case of MTTF substitute (3.17) into (3.16) to get

$$\text{MTTF} = \int_0^\infty t\lambda e^{-\lambda t} \qquad (3.20)$$

The following is obtained by integrating the above expression by parts:

$$\text{MTTF} = \left[-te^{-\lambda t} \right]_0^\infty - \left[-\frac{e^{-\lambda t}}{\lambda} \right]_0^\infty$$

$$\therefore \text{MTTF} = \frac{1}{\lambda} \qquad (3.21)$$

The above expression represents the situation when a component's failure times are exponentially distributed. MTTF is a reciprocal of the constant hazard rate, λ, as given by (3.21).

3.5 RELIABILITY NETWORKS

This section describes the five typical reliability configurations.

Figure 3.2   A series system block diagram.

3.5.1   Series Structure

This arrangement represents a system whose subsystems or components form a series network. If anyone of the subsystem or component fails, the series system experiences an overall system failure. A typical series system configuration is shown in Figure 3.2.

If the series system component failures are statistically independent, then the reliability R_s of a series system with nonidentical components is given by

$$R_s = \prod_{i=1}^{n} R_i \tag{3.22}$$

where n is the number of components or subsystems and R_i is the reliability of ith component or subsystem.

If the failure times of components are exponentially distributed (i.e., if components have constant failure rates), then the ith component reliability may be obtained from (3.19), that is,

$$R_i(t) = e^{-\lambda_i t} \tag{3.23}$$

By substituting (3.23) into (3.22),

$$R_s(t) = e^{-\sum_{i=1}^{n} \lambda_i t} \tag{3.24}$$

MTTF is given by

$$\text{MTTF} = \int_0^\infty e^{-\sum_{i=1}^{n} \lambda_i t} \, dt$$

$$= \frac{1}{\sum_{i=1}^{n} \lambda_i} \tag{3.25}$$

The above expression shows that a series system (MTTF) is the reciprocal of sum of the series network component failure rates.

Example 2. Two nonidentical pumps are required to run a system at a full load. Assume, pump I and II have constant failure rates $\lambda_1 = 0.0001$ failure/hour and $\lambda_2 = 0.0002$ failure/hour, respectively. Calculate this series system mean time to failure and reliability for a 100 hour mission; assume that both the pumps start operating at $t = 0$.

The following series system reliability R_s for a 100 hour mission is computed by using (3.24):

$$R_s(t) = e^{-(\lambda_1 + \lambda_2)t}$$

$$R_s(100) = e^{-(0.0001 + 0.0002)(100)} = 0.97045$$

By utilizing (3.25) we get

$$\text{MTTF} = \frac{1}{\lambda_1 + \lambda_2} = \frac{1}{0.0001 + 0.0002} = 3{,}333.3 \text{ hours}$$

3.5.2 Parallel Configuration

This configuration is shown in Figure 3.3. This system will fail if and only if all the units in the system malfunction. The model is based on the assumption that all the system units are active and load sharing. In addition it is assumed that the component failures are statistically independent. A parallel structure reliability R_p with nonidentical units or component reliability is given by

$$R_p = 1 - \prod_{i=1}^{n} (1 - R_i) \tag{3.26}$$

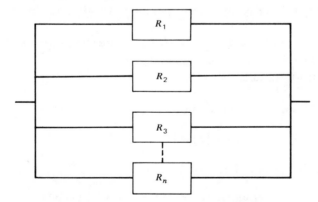

Figure 3.3 A parallel network block diagram.

where n is the number of units. R_i is the reliability of ith component or subsystem.

If the component failure rates are constant, then by substituting (3.19) into (3.26),

$$R_p(t) = 1 - \prod_{i=1}^{n} (1 - e^{-\lambda_i t}) \qquad (3.27)$$

MTTF is obtained by integrating (3.27) over the interval $[0, \infty]$,

$$MTTF = \int_0^\infty R_p(t)\, dt = \int_0^\infty \left\{ 1 - \prod_{i=1}^{n} (1 - e^{-\lambda_i t}) \right\} dt$$

$$= \left(\frac{1}{\lambda_1} + \frac{1}{\lambda_2} + \cdots + \frac{1}{\lambda_n} \right) - \left(\frac{1}{\lambda_1 + \lambda_2} + \frac{1}{\lambda_1 + \lambda_3} + \cdots \right)$$

$$+ \left(\frac{1}{\lambda_1 + \lambda_2 + \lambda_3} + \frac{1}{\lambda_1 + \lambda_2 + \lambda_4} + \cdots \right)$$

$$+ (-1)^{n+1} \frac{1}{\displaystyle\sum_{i=1}^{n} \lambda_i} \qquad (3.28)$$

For identical components, the above equation reduces to

$$MTTF = \frac{1}{\lambda} \sum_{i=1}^{n} \frac{1}{i} \qquad (3.29)$$

Example 3. Suppose two identical motors are operating in a redundant configuration. If either of the motors fails, the remaining motor can still operate at the full system load. Assume both motors are identical and their failure rates are constant. In addition, motor failures are statistically independent. If both motors start operating at $t = 0$, find the following:

1. System reliability for given $\lambda = 0.0005$ failure/hour, $t = 400$ hours (mission time).
2. Mean-time-to-failure (MTTF).

For identical units, (3.27) becomes

$$R(t) = 2e^{-\lambda t} - e^{-2\lambda t}$$

since $\lambda = 0.0005$ failure/hour, $t = 400$ hours

$$\therefore R(400) = 2e^{-(0.0005)(400)} - e^{-(2 \times 0.0005)(400)}$$

$$= 0.9671$$

MTTF is obtained from (3.29)

$$\text{MTTF} = \frac{1}{\lambda}\left(1 + \frac{1}{2}\right) = \frac{3}{2}\frac{1}{\lambda}$$

$$= \frac{1.5}{0.0005} = 3{,}000 \text{ hours}$$

3.5.3 Standby Redundancy

This type of redundancy represents a situation with one operating and n units as standbys. The standby redundancy arrangement is shown in Figure 3.4. Unlike a parallel network where all units in the configuration are active, the standby units are not active.

The system reliability of the $(n+1)$ unit, in which one unit is operating and n units on the standby mission until the operating unit fails, is given by

$$R_{st}(t) = \sum_{i=0}^{n} \frac{(\lambda t)^i e^{-\lambda t}}{i!} \tag{3.30}$$

The above equation is true if the following are true:

1. The switching arrangement is perfect.
2. The units are identical.
3. The units failure rates are constant.
4. The standby units are as good as new.
5. The unit failures are statistically independent.

In the case of $(n+1)$, nonidentical units whose failure time density functions are different, the standby redundant system failure density is

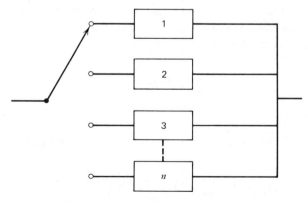

Figure 3.4 A standby redundancy model.

given by

$$f_{st}(t) = \int_{y_n=0}^{t} \int_{y_{n-1}=0}^{y_n} \cdots \int_{y_1=0}^{y_2} f_1(y_1) f_2(y_2 - y_1) \cdots f_{n+1}(t - y_n)$$

$$dy_1 \, dy_2 \cdots dy_n \tag{3.31}$$

Thus, the system reliability can be obtained by integrating $f_{st}(t)$ over the interval $[t, \infty]$ as follows:

$$R_{st}(t) = \int_t^\infty f_{st}(t) \, dt \tag{3.32}$$

Example 4. Assume that a system contains two identical units, that is, one operating and the other on the standby mission. Furthermore, the units failure rates are constant. In addition assume that the standby unit is as good as new at the beginning of its mission. Evaluate system reliability for a 100-hour mission for given unit failure rate, $\lambda = 0.001$ failure/hour.

Equation (3.30) gives us

$$R_{st}(t) = (1 + \lambda t) e^{-\lambda t} \tag{3.33}$$

For known $t = 100$ hours, $\lambda = 0.001$ failure/hour, the system reliability is

$$= (1 + 0.1) e^{-0.1} = 0.9953$$

3.5.4 k-*out of*-n *Configuration*

This is another form of redundancy. It is used where a specified number of units must be good for the system success. The series and parallel configuration in the preceding sections are special cases of this configuration, that is, $k = n$ and $k = 1$, respectively.

Reliability of this type of configuration is obtained by applying the binomial distribution. The system reliability for k-out-of-n number of independent and identical units is given by:

$$R_{k/n} = \sum_{i=k}^{n} \binom{n}{i} R^i (1 - R)^{n-i} \tag{3.34}$$

For the constant unit failure rate λ, the above equation becomes

$$R_{k/n}(t) = \sum_{i=k}^{n} \binom{n}{i} (e^{-\lambda t})^i \frac{(1 - e^{-\lambda t})^n}{(1 - e^{-\lambda t})^i} \tag{3.35}$$

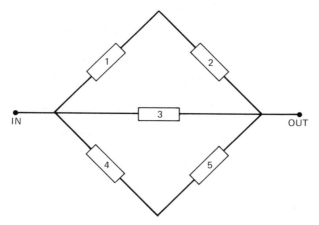

Figure 3.5 A five dissimilar units bridge network.

Example 5. Assume that in (3.35) $k=2$, $n=3$, and $\lambda=0.0001$ failure/hour. Therefore the system reliability for a 200-hour mission is

$$R_{2/3}(200)=3e^{-(2\times0.001)200}-2e^{-(3\times0.001)200}$$

$$=0.9133$$

3.5.5 Bridge Configuration

This network is shown in Figure 3.5. The critical element of the configuration is labeled as "3." For nonidentical and independent units, the five units bridge network reliability equation from reference 27 is

$$R_b=2R_1R_2R_3R_4R_5-R_2R_3R_4R_5-R_1R_3R_4R_5-R_1R_2R_4R_5$$

$$-R_1R_2R_3R_5-R_1R_2R_3R_4+R_1R_3R_5+R_2R_3R_4+R_1R_4+R_2R_5$$

$$(3.36)$$

In this case of identical units, the above equation reduces to

$$R_b=2R^5-5R^4+2R^3+2R^2 \tag{3.37}$$

For units with constant failure rate, substitute (3.19) into (3.37), that is,

$$R(t)=2e^{-5\lambda t}-5e^{-4\lambda t}+2e^{-3\lambda t}+2e^{-2\lambda t} \tag{3.38}$$

MTTF is obtained by integrating (3.38) over the interval $[0,\infty]$, that is,

$$\text{MTTF}=\int_0^\infty (2e^{-5\lambda t}-5e^{-4\lambda t}+2e^{-3\lambda t}+2e^{-2\lambda t})\,dt$$

$$=\frac{49}{60}\frac{1}{\lambda} \tag{3.39}$$

Example 6. Compute system reliability and MTTF for a five independent and identical units bridge configuration. Suppose

$$\lambda = 0.0005 \text{ failure/hour}$$

$$t = 100 \text{ hours}$$

and all units start operating at $t = 0$. Equation 3.38 gives

$$R(300) = 2e^{-0.25} - 5e^{-0.2} + 2e^{-0.15} + 2e^{-0.1}$$

$$= 0.9999$$

The following MTTF result is obtained by substituting the given failure data in (3.39)

$$\text{MTTF} = \frac{49}{60 \times 0.0005} = 1633.4 \text{ hours}$$

3.6 RELIABILITY EVALUATION TECHNIQUES

This section briefly describes reliability evaluation techniques Readers with no indepth knowledge of these techniques should consult references 36 and 37.

3.6.1 *Binomial Theorem to Evaluate Network Reliability*

This is one of the simplest methods to evaluate system reliability. However, it is only useful for evaluating reliability of simple systems of series and parallel form. For complex systems it is quite a trying task.

The following is always true for the binomial expression in reliability engineering:

$$(p+q)^n = 1 \qquad (3.40)$$

where P = the component probability of success
q = the component probability of failure
n = the number of identical components

Example 7. When two identical components form a series or parallel configuration we obtain the resulting equation from (3.40):

$$(p+q)^2 = p^2 + 2pq + q^2 \qquad (3.41)$$

Here $p^2 \equiv$ probability of both components operating
$2pq \equiv$ probability of one component failed and one working
$q^2 \equiv$ probability of both components failed

Therefore, the reliability equation of a two unit parallel system is given by the first two terms of the right-hand side of expression (3.41):

$$R = p^2 + 2pq \tag{3.42}$$

$$\text{since } p + q = 1, \text{ then } p = 1 - q \tag{3.43}$$

By substituting (3.43) into (3.42) we get

$$R = 1 - q^2 \tag{3.44}$$

This is the reliability equation for a two identical and independent unit parallel system.

3.6.2 State Space Approach (Markov Processes)

The state space approach is a very general approach and can generally handle more cases than any other method. It can be used when the components are independent as well as for systems involving dependent failure and repair modes. There is no conceptual difficulty in incorporating multistate components and modeling common cause failures.

The method proceeds by the enumeration of system states. The state probabilities are then calculated and the steady-state reliability measures can be calculated using the frequency balancing approach [39]. The pertinent relationships are given below:

1. Unavailability or the probability of failure is given by

$$P_f = \sum_{i \in F} p_i \tag{3.45}$$

where p_i = probability of being in state i
F = subset of failure states

2. Frequency of failure [39] of encountering subset F is given by

$$f_f = \sum_{i \in S-F} p_i \sum_{j \in F} \lambda_{ij} \tag{3.46}$$

where S = system state space
λ_{ij} = transition rate from state i to state j

3. Mean duration of failure state is [39]

$$d_f = \frac{P_f}{f_f} \tag{3.47}$$

When the components are independent, the state probabilities can be obtained using the multiplication rule. When, however, dependent failure or repair modes are involved, the state probabilities have to be obtained by solving a set of linear algebraic equations. Probably the only serious problem, particularly when constant transition rates are assumed, with this approach is that for large systems, it could become unmanageable. In many situations, the problem can be handled using a computer-generated state transition matrix and reducing the size of the state space by truncation, sequential truncation, and using the concept of state merging. These techniques are discussed in reference 39. Examples of the application of these techniques for large systems can be found in reference 40.

3.6.3   Network Reduction Technique

It is a simple and useful procedure for systems consisting of series and parallel subsystems. Configurations such as bridge networks can be analyzed using delta-star conversions [12, 13, 41]. Some approximation, however, is involved in the use of these techniques [41]. The technique consists of sequentially reducing the parallel and series configurations to equivalent units until the whole network reduces to a single unit. The bridge configurations can be converted to series and parallel equivalents by using delta-star conversions or decomposition approach.

The primary advantage of this method is that it is easy to understand and apply; however, generally it is not suitable for considering degraded failure modes of components and systems. The independence of components has to be generally assumed.

3.6.4   Decomposition Method

This method decomposes a complex system into simpler subsystems by the application of conditional probability and conditional frequency theorems [37]. The reliability measures of simpler subsystems are calculated and then combined to obtain the results for the system. This method can be used to simplify both the state space, as well as the network approach. Examples of application in both of these areas can be found in reference 37. The success of the method depends upon the choice of the key component, that is, the component used for decomposing the network. If this component is not judiciously chosen, the final results will be the same, but the computations could be far more tedious. For a relatively complex network, the choice of proper key components to decompose the system into series parallel configurations can be a trying task.

3.6.5   Minimal Cut Set Method

A general approach to the solution of reliability block diagrams is based on minimal cut sets or minimal tie sets. This approach is very suitable for computer application. The minimal cut sets of a reliability block diagram can be identified using special algorithms. Once the minimal cut sets have been enumerated, the reliability measures can be calculated using the following relationships:

$$P_f = P(\overline{C}_1) + \cdots + P(\overline{C}_m) - \left[P(\overline{C}_1 \cap \overline{C}_2) + \cdots + P\left(\underset{i \ne j}{\overline{C}_i \cap \overline{C}_j} \right) \right]$$

$$\cdots (-1)^{m-1} \left[P(\overline{C}_1 \cap \cdots \cap \overline{C}_m) \right] \tag{3.48}$$

and

$$f_f = P(\overline{C}_1)\bar{\mu}_1 + \cdots + P(\overline{C}_m)\bar{\mu}_m - P(\overline{C}_1 \cap \overline{C}_2)\bar{\mu}_{1+2}$$

$$+ \cdots + P\left(\underset{i \ne j}{\overline{C}_i \cap \overline{C}_j} \right)\bar{\mu}_{1+j} \Bigg)$$

$$\cdots (-1)^{m-1} \left[P(\overline{C}_1 \cap \cdots \cap \overline{C}_m)\bar{\mu}_{1+\cdots+m} \right) \tag{3.49}$$

where $C_i, \overline{C}_i$ = the minimal cut set i and failure of components in C_i, respectively

μ_i = the repair rate of component i

$\bar{\mu}_{i+j+k}$ = the sum of μ_j over all $j \in (C_i \cup C_j \cup C_k)$, that is, the sum of repair rates of the components which belong to any or all of C_i, C_j, C_k

P_f = probability of failure

f_f = frequency of failure

As in the case of all network methods, this technique is not suitable for incorporating degraded modes of operation. For m minimal cut sets, the number of terms to be evaluated is 2^m. This could create computational problems, which could be partly alleviated using the concept of probability and frequency bounds [42]. Similarly, one may obtain tie sets for a complex system (described in detail in reference 37).

3.6.6 Delta-Star Technique

To analyze a complex structure such as bridge, the delta-star transformation [12, 13] easily transforms the configuration into series and parallel combinations. We derive the delta-star equivalent formulas by obtaining the equivalent legs of the block diagrams of Figure 3.6.

Consider, for example, three components of a system with reliabilities R_{AC}, R_{AB}, and R_{CB} connected to form the delta configuration shown in

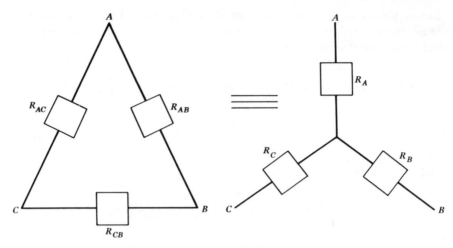

Figure 3.6 Delta-star reliability equivalent.

Figure 3.6. This configuration yields the star equivalent with reliabilities R_A, R_B and R_C.

Now consider the transformation steps indicated by Figure 3.7 to derive the delta to star equivalent. The application of independent event probability laws to components connected in series and parallel combinations as shown in Figure 3.7(a–c) will yield (3.55), (3.56), and (3.57), respectively:

For a simple independent series, the total system reliability is given by

$$R_T = R_A R_B \cdots R_N \tag{3.50}$$

where R_A, $R_B \cdots R_N$ are the reliabilities of the N components.

The simple independent parallel case yields the total system failure probability as

$$F_T = F_A F_B \cdots F_N = (1 - R_A)(1 - R_B) \cdots (1 - R_N) \tag{3.51}$$

where F_A, $F_B \cdots F_N$ are the unreliabilities of the N components.

Applying (3.50) and (3.51) to the legs presented in Figure 3.7 their corresponding relationships are

$$R_A R_B = 1 - (1 - R_{AB})(1 - R_{AC} R_{CB}) \tag{3.52}$$

$$R_B R_C = 1 - (1 - R_{CB})(1 - R_{AC} R_{AB}) \tag{3.53}$$

$$R_A R_C = 1 - (1 - R_{AC})(1 - R_{CB} R_{AB}) \tag{3.54}$$

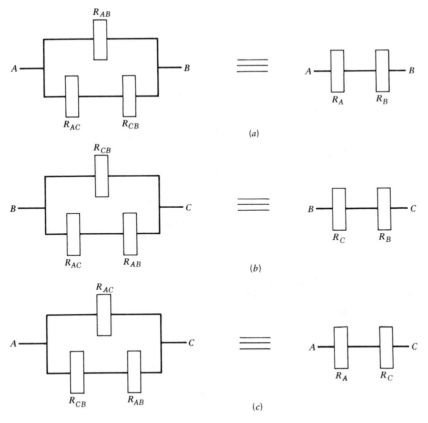

Figure 3.7 Delta-star equivalent legs.

From these three simultaneous equations the following delta-to-star relationships result:

$$R_A = \sqrt{\frac{[1-(1-R_{AC})(1-R_{CB}R_{AB})][1-(1-R_{CB})(1-R_{AC}R_{AB})]}{[1-(1-R_{AB})(1-R_{AC}R_{CB})]}}$$

$$(3.55)$$

$$R_B = \sqrt{\frac{[1-(1-R_{AB})(1-R_{AC}R_{CB})][1-(1-R_{CB})(1-R_{AC}R_{AB})]}{[1-(1-R_{AC})(1-R_{CB}R_{AB})]}}$$

$$(3.56)$$

$$R_C = \sqrt{\frac{[1-(1-R_{AC})(1-R_{CB}R_{AB})][1-(1-R_{AB})(1-R_{AC}R_{CB})]}{[1-(1-R_{CB})(1-R_{AC}R_{AB})]}}$$

$$(3.57)$$

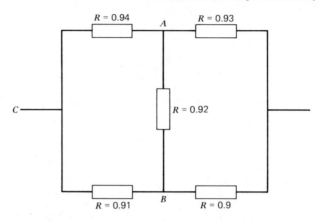

Figure 3.8   A two-state device bridge structure.

Example 8.   A bridge network example for independent units is solved to illustrate the use of these formulas. Figure 3.8 illustrates the structure for a simple bridge; the letters A, B, and C are used to label elements of the delta configuration.

We obtain the equivalent star configuration values by using (3.55), (3.56), and (3.57).

$$\therefore R_A = 0.9948$$

$$R_B = 0.9930, \; R_C = 0.9954$$

The network shown in Figure 3.8 may be expressed as its equivalent as shown in Figure 3.9. The reliability equation for this structure is

$$R_T = \left[1 - (1 - R_1 R_A)(1 - R_2 R_B) \right] R_C$$

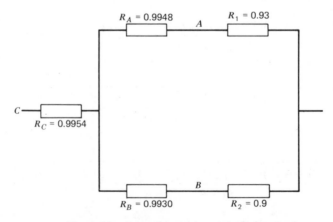

Figure 3.9   A transformed two-state device structure.

Numerically the value of the total bridge reliability is

$$R_T = 0.987$$

for the given component reliability values.

Analyzing this bridge structure with the event space method also yields

$$R_T = 0.987$$

Equations 3.55–3.57 are interrelated. Therefore computation of the value of the first equation helps to compute values of the other two equations. This minimizes the computing time.

3.7 FAILURE MODE AND EFFECT ANALYSIS (FMEA)

This is an important step in a reliability and maintainability assurance program. FMEA is a tool to evaluate design at the initial stage from the reliability aspect. This criteria helps to identify need for and the effects of design change.

Furthermore, the procedure demands listing of potential failure modes of each and every component on paper and its effects on the listed subsystems.

FMEA becomes failure modes, effects, and criticality analysis (FMECA) if criticalities or priorities are assigned to failure mode effects.

Some of the main characteristics of this procedure are as follows:

1. This is a routine upward procedure that begins from the detailed level.
2. By evaluating failure effects of each component, the whole system is screened completely.
3. It improves communication among design interface personnel.
4. It identifies weak areas in a system design and indicates areas where further or detailed analysis are desirable.

Some of the main steps to perform FMEA are shown in Figure 3.10.

3.8 RELIABILITY APPORTIONMENT

To achieve the required reliability of a complex system, it is a routine procedure to set reliability targets for subsystems. Its main advantage is that once the individual subsystem reliability goal is achieved, then the overall system goal will automatically be fulfilled.

The process to set such reliability goals is known as the reliability apportionment. Normally this is accomplished before the key design or development decisions are made.

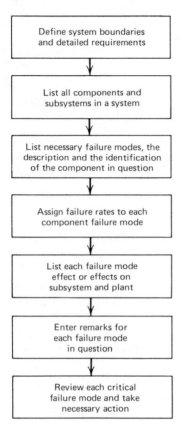

Figure 3.10   FMEA flow chart.

Some of the reliability apportionment techniques are described in the following sections.

3.8.1   Reliability/Cost Models

Before applying this reliability/cost procedure, the relationship between reliability and cost must be known for each subsystem to meet system reliability goal at minimum cost. However the main drawback of this procedure is the lack of availability of cost data, that is, cost at a given level of reliability.

3.8.2   Similar Familiar Systems Reliability Apportionment Approach

This approach is based upon the familiarity of the designer with similar systems or subsystems. Its main weakness stems from the fact that the reliability and life cycle cost of earlier similar designs have to be assumed adequate when designing new systems. By applying this technique the failure data collected on similar systems from the various sources can be utilized.

3.8.3 Factors of Influence Method

This procedure is purely based upon the following important factors that effect the system in question:

1. *Complexity/Time.* In the case of complexity it relates to the number of subsystem parts, whereas time is related to the relative operational time during the total functional period.
2. *Environmental Factor.* This concerns each subsystem's operating environmental conditions such as temperature, vibration, humidity. In other words it deals with susceptibility or exposure of subsystems to such environmental conditions.
3. *State-of-the-Art.* This factor takes care of advancement in the state-of-the-art for a particular subsystem or component.
4. *Subsystem Failure Criticality.* This factor includes the criticality effect of a subsystem failure on the system. For example, the failure of some auxiliary instruments in an aircraft may not be as critical as the failure of engine.

When applying this factor of influence procedure, each and every subsystem is rated with respect to the influential factors, and one can assign a number between 1 and 10, where 1 is allocated to a subsystem least affected by the factor in question and 10 is allocated to a subsystem most affected by the factors of influence. Thus reliability can be allocated by using the weight of these assigned numbers for all factors.

3.8.4 Combined Familiar Systems and Factor of Influence Method

Both the familiar systems and factors of influence methods have their weakness when they are used individually. However, combining the two methods produces better results because data are used from the similar subsystems as well as when new subsystems are designed under different factors of influence.

REFERENCES

1. Amstadter, B. L., *Reliability Mathematics*, McGraw-Hill, New York, 1971.
2. Von Alven, W. H. (Ed.), *Reliability Engineering*, Prentice-Hall, Englewood Cliffs, NJ, 1964.
3. Barlow, R. E. and F. Proschan, *Mathematical Theory of Reliability*, Wiley, New York, 1965.

4. Barlow, R. E. and F. Proschan, *Statistical Theory of Reliability and Life Testing*, Holt, Rinehart and Winston, New York, 1975.

5. Bazovsky, I., "Fault Trees, Block Diagrams and Markov Graphs," *Proceedings of the* 1977 *Annual Reliability and Maintainability Symposium*, *Philadelphia*, IEEE, New York, 1977, pp. 134–141.

6. Bazovsky, I., *Reliability: Theory and Practice*, Prentice-Hall, Englewood Cliffs, NJ, 1961.

7. Billinton, R., *Power-System Reliability Evaluation*, Gordon and Breach, New York, 1970.

8. Billinton, R., R. J. Ringlee, and A. J. Woods, *Power-System Reliability Calculations*, MIT Press, Cambridge, MA, 1973.

9. Blanchard, B. S., Jr. and E. E. Lawery, *Maintainability*, McGraw Hill, New York, 1969.

10. Calabro, S. R., *Reliability Principles and Practice*, McGraw-Hill, New York, 1962.

11. Cox, D. R., *Renewal Theory*, Methuen, London, 1962.

12. Dhillon, B. S., The Analysis of the Reliability of Multi-State Device Networks, Ph.D. Thesis (1975). (Available from National Library of Canada, Ottawa.)

13. Singh (Dhillon), B., and C. L. Proctor, "Reliability Analysis of Multi-State Device Networks." *Proceedings of the* 1976 *Annual Reliability and Maintainability Symposium*, *Las Vegas*, IEEE, New York, 1976, pp. 31–35.

14. Dhillon, B. S., and C. Singh, "Bibliography of Literature on Fault Trees," *Microelectr. Reliability*, **18**, 501–503 (1978).

15. Dhillon, B. S., "A Hazard Rate Model," *IEEE Trans. Reliability* **R-28**, pp. 150, (1979).

16. Epstein, B., *Mathematical Models for System Reliability*, Student Association Technion, Israel Institute of Technology, Haifa, Israel, 1969.

17. Gnedenko, B. V., Y. K. Belyaev, and A. D. Solovyev, *Mathematical Methods of Reliability Theory*, Academic, New York, 1969.

18. Goldman, A. S., and T. B. Slattery, *Maintainability*, Wiley, New York, 1964.

19. Grouchko, D., (Ed.), *Operations Research and Reliability*, Gordon and Breach, New York, 1969.

20. Haviland, R. T., *Engineering Reliability and Long-Life Design*, Van Nostrand, Princeton, NJ, 1964.

21. Howard, R. A., *Dynamic Probabilistic Systems*, Vols. I and II, Wiley, New York, 1971.

22. Ireson, W. G., (Ed.), *Reliability Handbook*, McGraw-Hill, New York, 1966.

23. Jardine, A. K. S., (Ed.), *Operational Research in Maintenance*, Manchester University Press, Manchester and Barnes and Nobles, New York, 1970.

24. Jorgenson, D. W., J. J. McCall, and R. Radner, *Optimal Replacement Policy*, Rand McNally, Chicago, 1967.

25. Kapur, K. C., and L. R. Lamberson, *Reliability in Engineering Design*, Wiley, New York, 1977.

26. Kozlov, B. A., and I. A. Ushakov, *Reliability Handbook*, Holt, Rinehart and Winston, New York, 1970.

27. Lipp, J. P., "Topology of Switching Elements vs Reliability," *Trans. IRE Reliability Qual. Control*, 7, 21–34 (June 1957).

28. Mann, N. R., R. E. Shafer, and N. D. Singpurwalla, *Methods for Statistical Analysis of Reliability and Life Data*, Wiley, New York, 1974.

29. Morse, P. M., *Queues, Inventories and Maintenance*, Wiley, New York, 1958.

30. Myer, R. H., K. L. Wong, and H. M. Gordy (Ed.), *Reliability Engineering for Electronic Systems*, Wiley, New York, 1964.

31. Pieruschaka, E., *Principles of Reliability*, Prentice-Hall, Englewood Cliffs, NJ, 1963.

32. Polovko, A. M., *Fundamentals of Reliability Theory,* Academic, New York, 1968.

33. *Proceedings of the Annual Reliability and Maintainability Symposia* and its predecessors (available from Annual Reliability and Maintainability Symposium; 6411 Chillum Place NW, Washington, D.C. 20012; or from the IEEE).

34. Roberts, N. H., *Mathematical Methods in Reliability Engineering*, McGraw-Hill, New York, 1965.

35. Sandler, G. H., *System Reliability Engineering*, Englewood Cliffs, NJ, Prentice-Hall, 1963.

36. Shooman, M. L., *Probabilistic Reliability: An Engineering Approach*, McGraw-Hill, New York, 1968.

37. Singh, C., and R. Billinton, *System Reliability Modeling and Evaluation*, Hutchinson, London (1977).

38. Singh, C., *Calculating the Frequency of Boolean Expressing Being 1, IEEE Trans. Reliability,* **R-26**, 354–355 (1977).

39. Singh, C., "Reliability Calculations on Large Systems," *Proceedings of 1975 Annual Reliability and Maintainability Symposium, Washington, D.C.*, IEEE, New York, 1975, pp. 188–193.

40. Singh, C., "Reliability Models for Track Bound Transit Systems," *Proceedings of the 1977 Annual Reliability and Maintainability Symposium, Philadelphia*, IEEE, New York, 1977, pp. 242–247.

41. Singh, C., and M. D. Kankam, "Comments on Closed Form Solutions for Delta-Star and Star-Delta Conversion of Reliability Networks," *IEEE Trans. Reliability*, **R-25**, 336–339 (1976).

42. Singh, C., *On the Behaviour of Failure Frequency Bounds, IEEE Trans. Reliability*, **R-26**, 63–66 (1977).

43. Smith, C. O., *Introduction to Reliability in Design*, McGraw-Hill, New York, (1976).

44. Smith, D. J., *Reliability Engineering*, Pitman, New York, 1973.

45. Smith, D. J., and A. H. Babb, *Maintainability Engineering*, Pitman, New York, 1973.

46. Tsokos, C. P., and I. N. Shimi (Ed.), *The Theory and Applications of Reliability: With Emphasis on Bayesian and Nonparametric Methods*, Academic, New York (1977).

47. Vogt, W. G., and M. H. Mickle (Ed.), *Modeling and Simulation Conference Proceedings*, Instrument Society of America, 1969–1979.

48. Zelen, M., (Ed.), *Statistical Theory of Reliability*, Madison, WI, University of Wisconsin Press, 1963.

4

Fault Trees and Common Cause Failures

4.1 INTRODUCTION

This chapter discusses fault trees, which are to analyze complex systems. This technique has rapidly gained favor because of its versatility in degree of detail of complex systems. The fault tree technique was originated by H. A. Watson of Bell Telephone Laboratories to analyze the Minuteman Launch Control System. It was further refined by a study team at the Bell Telephone Laboratories.

Further work on fault tree techniques was carried out at the Boeing company in which Haasl [37] played an instrumental role. A turning point took place in 1965 when several papers on the technique were presented at the 1965 Safety Symposium held at the University of Washington, Seattle, [37]. Ever since several experts have made further advances in this technique.

Again another symposium on the technique was organized at the University of California at Berkeley [2]. A comprehensive bibliography on the technique is presented in reference 21.

Most of the material presented in this chapter is taken from the listed fault tree bibliography at the end of this chapter. The second part of this chapter deals with the subject of common-cause failures.

4.2 FAULT TREE SYMBOLS AND DEFINITIONS

This section presents most of commonly used fault tree symbols and definitions. For more comprehensive symbols and definitions one should consult references 65 and 124.

AND Gate. The AND gate denotes that an output event occurs if and only if all the input events occur.

OR Gate.   The OR gate denotes that an output event occurs if any one or more of the input events occur.

Exclusive OR Gate.   The output of this gate is an intermediate event. This gate denotes that there is no output unless one and only one of the input events occurs.

Priority AND Gate.   It is logically equivalent to an AND gate with the exception that the input events must occur in a specific order. It is represented by the following symbol:

Inhibit Gate.   This gate produces output only when the conditional input is satisfied. The inhibit gate is logically equivalent to an "AND" gate with two input events.

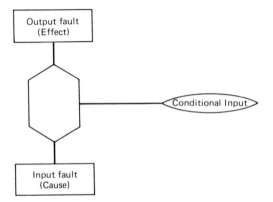

Special Gate.   This gate represents any other legitimate combination of the input events.

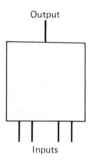

Delay Gate.   This represents a gate whose output only occurs after a specified delay time has elapsed.

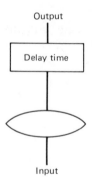

The Triangle.   A triangle denotes a transfer IN or OUT. It is used to avoid repeating sections of the fault tree. A line from the top of the triangle indicates "transfer in." A line from the side of the triangle denotes "transfer out."

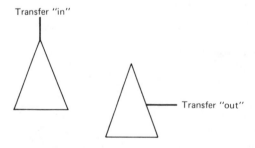

Resultant Event. A rectangle denotes an event which results from the combination of fault events through the input of a logic gate.

Basic Fault Event. A circle represents a basic fault event or the failure of an elementary component. The failure parameters such as unavailability, probability, failure, and repair rates of a fault event are obtained from the empirical date or other sources.

Incomplete Event. A diamond represents a fault event whose causes have not been fully developed. This event could be further developed to show basic contributary failures; however, it is not developed either due to lack of information or due to lack of interest.

Trigger Event. The house shape symbol denotes a fault event which is expected to occur.

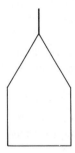

The Conditional Event. This is denoted by an ellipse. This symbol indicates any condition or restriction that applies to a logic gate.

Double Diamond Event. This symbol represents an undeveloped fault event that requires further development to accomplish the fault tree.

The Upside Down Triangle. This symbol denotes a similarity transfer, that is, the input is similar but not identical to the like identified input.

4.3 GENERAL PROCEDURE TO ANALYZE FAULT TREES

To develop fault trees, the following basic steps are generally required:

1. Define the undesired event (top event) of the system under consideration.
2. Thoroughly understand the system and its intended use.
3. To obtain the predefined system fault condition cause, determine the higher order functional events. In addition, continue the fault event analysis to determine the logical interrelationship of lower level events that can cause them.
4. After accomplishing steps 1–3 construct a fault tree of logical relationships among input fault events. These are to be defined in terms of basic, identifiable, and independent faults.

To obtain quantitative results for the top event (undesired event) assign failure probability, unavailability, failure, and repair rates data to basic events provided the fault tree events are redundancy free.

A more rigorous and systematic approach requires the following steps:

1. System definition.
2. Fault tree construction.
3. Qualitative evaluation.
4. Quantitative evaluation.

The above steps are outlined in detail in the following sections:

4.3.1 System Definition

To establish the system definition in fault tree analysis is a very difficult task. A system is normally represented by a functional layout diagram showing all functional interconnections and components of the system in question. To draw a fault tree of a system, it is strongly recommended that the system boundary conditions be established. However, care must be taken so that these boundary conditions are not confused with the physical bounds of the system.

One of the most important boundary requirements is the top event (undesired event). Therefore, care must be taken to define the system top event for which the fault tree is to be drawn, because this is a major system failure. In addition to make the fault tree analysis understandable to others, the analyst must list all the assumptions on system definition and fault tree.

4.3.2 Fault Tree Construction

The major objective of fault tree construction is to represent system conditions symbolically, which may cause the system to fail. Furthermore, the fault tree construction can pinpoint the system weaknesses in a visible form. This acts as a visual tool in communicating and supporting decisions based on the analysis and to perform trade off studies or determine the adequacy of the system design.

Generally the analyst is expected to understand the system thoroughly before he proceeds to construct a system fault tree. To enhance the fault tree analysis, a system description should be part of the analysis documentation.

There are three generally accepted approaches to construct fault trees:

1. Primary failure technique.
2. Secondary failure technique.
3. Commanded failure technique.

The above techniques are used at the discretion of the reliability analyst according to the main requirements of the failure fault tree analysis.

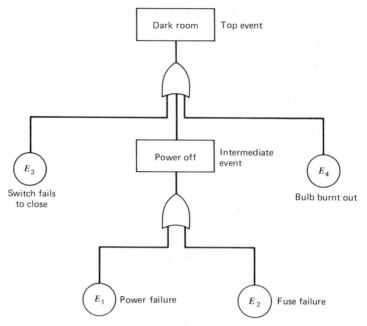

Figure 4.1   A primary failure fault tree.

Primary Failure Fault Tree Construction.   The failure of a component is called primary failure if it occurs while the part is functioning within the operating parameters for which it was designed. To construct a fault tree by only using primary failures is a straightforward process, because a fault tree is only developed to the point where identifiable primary component failures will produce fault events. The following example is presented to illustrate this technique.

Example 1.   Construct a fault tree of a simple system concerning a room containing a switch and a light bulb. Assume the switch only fails to close. In addition, the top event is the dark room.

The system fault tree is shown in Figure 4.1. The basic or primary events of the fault tree are as follows:

1. Power failure, E_1.
2. Fuse failure, E_2.
3. Switch fails to close, E_3.
4. Bulb burnt out, E_4.

The intermediate event is the "power off." The failure event of main concern is the top event, labeled "dark room." Therefore, the major emphasis of this analysis is toward the darkness in the room. The fault tree in Figure 4.1 shows that the input events are gated through OR gates. At the occurrence of any one of the basic four events E_1, E_2, E_3, E_4, the system top event ("dark room") will occur.

Secondary Failure Fault Tree Construction. To include secondary failures in fault tree analysis requires a greater insight into the system. The fault tree analysis is carried out beyond the basic component failure level. The secondary failures are due to excessive environmental or operational stress placed on the system components.

Example 2. A simple fault tree with the top event "motor fails to deliver power" is shown in Figure 4.2. The fault tree shows the primary events such as switch fails to close, internal motor circuitry failure, power failure,

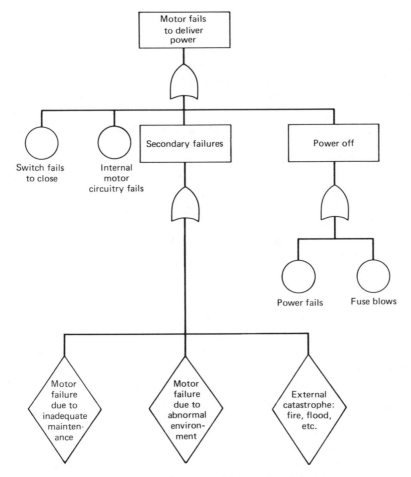

Figure 4.2 A fault tree with secondary failures.

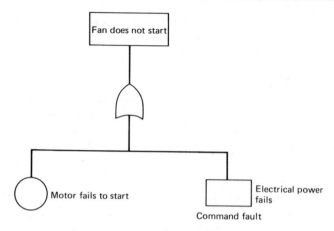

Figure 4.3   A basic and command failure fault tree.

and fuse failure. Secondary failures are represented in the rectangle as an intermediate event. The secondary failures shown in Figure 4.2 occur due to inadequate maintenance, hostile external environment, external catastrophe, and so on. These failures are discussed later in this chapter.

Fault Tree Construction with Command Failures.   These failures result from proper component operation at the wrong time or place. Command failures are failures of the coordinating events between various levels of the fault tree from basic failure events to the top event (undesired event or final event). A typical example of command failure is an erroneous electrical signal to an electrical device (e.g., a motor, a transducer). Figure 4.3 shows the interrelationship among basic and commanded failures. In Figure 4.3 the basic failure is represented by a circle, whereas the rectangle represents a commanded fault.

4.3.3   Qualitative Fault Tree Evaluation

This approach uses minimal cut sets of a fault tree. A cut set is defined as a set of basic events whose occurrence results in an undesired event. Furthermore, if a cut set cannot be reduced but insures the occurrence of the undesired event, the set is a minimal cut set. Obtaining minimal cut sets is a tedious process, since a computerized algorithm is required to obtain minimal cut sets. A qualitative evaluation example is presented in Figure 4.4.

As shown in Figure 4.4, the intermediate fault event B can only occur if both events E_1 and E_2 occur. In the case of intermediate event C, it can only occur if either event E_3 or event E_4 is present. The top event results if either one of the intermediate event B or C occurs.

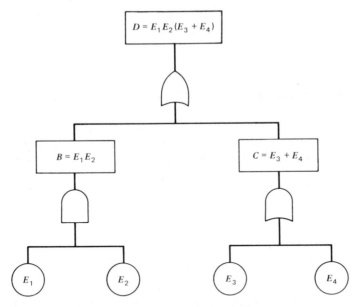

Figure 4.4 A hypothetical event fault tree.

4.3.4 Quantitative Fault Tree Evaluation

This evaluation uses top event quantitative reliability information, such as failure probability, failure rate, or repair rate. Component failure parameters are evaluated first, then critical path, and finally the top event.

There are two accepted methods to determine quantitative fault tree results:

1. The Monte Carlo simulation method.
2. The analytical solution approach.

In the case of the Monte Carlo simulation, the fault tree is simulated using a digital computer to obtain quantitative results. Generally, the fault tree failures are simulated over thousands or millions of trial years of performance. Some of the main steps required to simulate a fault tree on a digital computer are as follows:

1. Assign failure data to the basic events.
2. Represent the entire fault tree on a digital computer.
3. List failures that lead to occurrence of the top event and the associated minimal cut sets.
4. Compute the desired end results.

In the case of the direct analytical solution method, it makes use of the existing analytical techniques. These techniques are described in the forthcoming sections.

4.4 ANALYTICAL DEVELOPMENTS OF BASIC GATES

Fault trees are constructed to show system components pictorially and logically. A fault tree is constructed by using AND, OR, and other gates that relate logically various basic component faults to the top event. To represent these logic diagrams in a mathematical form, the Boolean algebra is an invaluable tool. The mathematical expressions for OR, AND, and Priority AND gates are developed in the following sections.

4.4.1 OR Gate

The OR gate is represented by the symbols U or $+$. Any one of these symbols denotes the union of events associated with an OR gate. A mathematical representation of two inputs OR gate is shown in Figure 4.5. The output event B_0 of an OR gate in Boolean algebra is written as

$$B_0 = B_1 + B_2 \tag{4.1}$$

where B_1 and B_2 are the input events.

4.4.2 AND Gate

In Boolean algebra the AND situation is represented by the symbol $\cdot$ or $\cap$. This symbol represents intersection of events. The two-input AND gate is shown in Figure 4.6. The output event, B_0, of the AND gate in Boolean algebra is represented by (4.2):

$$B_0 = B_1 \cdot B_2 \tag{4.2}$$

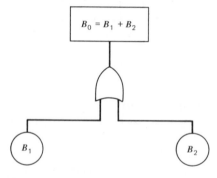

Figure 4.5 An OR gate with two input events.

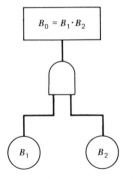

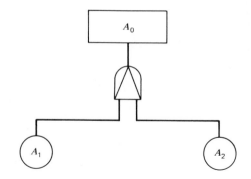

Figure 4.6 A two input AND gate.

Figure 4.7 A two input priority AND gate.

4.4.3 Priority AND Gate

This is logically equivalent to an AND gate with an exception that its input events must occur in a specified order. A two input priority AND gate is shown in Figure 4.7.

In this situation it is supposed that the event A_1 must occur before event A_2. The development of a mathematical expression for the gate is presented in reference 31.

4.5 A FAULT TREE WITH REPEATED EVENTS

This type of situation is illustrated in Figure 4.8. The alphabetic letters in the diagram represent the fault events; A_1, A_2, A_3, and C indicate the basic fault events; B_1, B_2, B_0, the mean intermediate fault events; T the top event.

The fault tree shown in Figure 4.8 can be represented by the Boolean expressions as follows:

$$T = C \cdot B_0 \tag{4.3}$$

$$B_0 = B_1 \cdot B_2 \tag{4.4}$$

$$B_1 = (A_1 + A_2) \tag{4.5}$$

$$B_2 = (A_1 + A_3) \tag{4.6}$$

By substituting expressions (4.4) and (4.6) in expression (4.3) we get

$$T = C \cdot (A_1 + A_2) \cdot (A_1 + A_3) \tag{4.7}$$

It is clearly shown in Figure 4.8 that the event A_1 is the repeated basic fault event. Therefore, the expression (4.7) has to be simplified by applying the basic Boolean algebra properties.

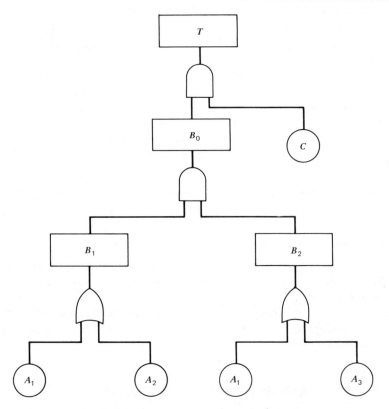

Figure 4.8 A fault tree with repeated events.

Basic Boolean Algebra Properties

1. Laws of absorption:

$$A+(A\cdot B)=A \tag{4.8}$$

$$A\cdot(A\cdot B)=A\cdot B \tag{4.9}$$

2. Identities:

$$A+A=A \tag{4.10}$$

$$A\cdot A=A \tag{4.11}$$

3. Distributive laws:

$$A+B\cdot C=(A+B)(A+C) \tag{4.12}$$

By applying distributive law of expression (4.12) to expression (4.4) we get

$$B_0=A_1+A_2\cdot A_3 \tag{4.13}$$

By using expressions (4.10) and (4.11) in (4.3), expression (4.7) reduces to

$$T=C\left[\,A_1+A_2\cdot A_3\right] \tag{4.14}$$

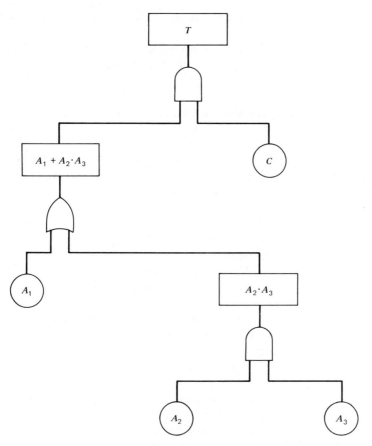

Figure 4.9 A simplified fault tree.

Because of expression (4.14) our original fault tree of Figure 4.8 reduces to the one shown in Figure 4.9.

Therefore, it is always recommended to reduce the repeated event expression by applying the Boolean properties before obtaining the quantitative reliability parameter results. Otherwise, the quantitative results will be misleading. Algorithms to obtain repeated events free fault tree are presented in references 1, 27, 33, 66, and 12. One such algorithm is presented in the section to follow:

4.6   AN ALGORITHM TO OBTAIN MINIMAL CUT SETS

A difficult problem associated with the fault tree technique is to obtain minimal cut sets of a fault tree. Here we present an algorithm developed in references 33, 27, and 1. Other computer oriented algorithms may be found in references 66, 12, and 9.

Before we present this algorithm we would like to present the definitions of both cut set and minimal cut set. These definitions are taken from reference 1.

A Cut Set.   This is a collection of basic events whose presence will cause the top event to occur.

A Minimal Cut Set.   A cut set is said to be minimal if it cannot be further minimized but still insures the occurrence of the top event. Minimal cut sets are sometimes called the minimal failure modes of a system.

The algorithm presented here can be used manually for simple fault trees. However, for a complex fault tree with hundreds of gates and basic events, it has to be computerized.

The algorithm is quite efficient. The main features of this algorithm are that the AND gate always increases the size of a cut set, whereas an OR gate increases the number of cut sets. These facts will be self-explanatory in the following solved example. We thought a solved example will be more useful to understand the practical aspect of this algorithm rather than presenting the background theory on the topic. Therefore, the readers interested in the theoretical background should consult reference 33.

Example 3.   The fault tree of the hypothetical example is shown in Figure 4.10. The gates are labeled as GT and the basic events as numerals. This algorithm begins from the gate below the top event in the example. It is labeled as GT0. As we know from our past basic knowledge on fault trees, the top event gate may normally be AND or OR gate.

However, if the top event gate, GT0, is an OR gate then each input to the OR gate represents an entry for each row of the list matrix. Whereas, in the case of an AND gate, each input represents an entry for each column of the list matrix.

For example, as shown in Figure 4.10, the top event gate, GT0, is an OR gate, therefore, we begin the formulation of the list matrix by listing inputs, GT1 and GT2 (output events) in a single column but in separate rows as follows:

$$
\boxed{\begin{array}{c} 1 \\ GT1 \\ GT2 \end{array}} \; —Step\ 1
$$

Any one input of an OR gate will cause the occurrence of an output event. Therefore, the inputs of the GT0 are the members of separate cut sets.

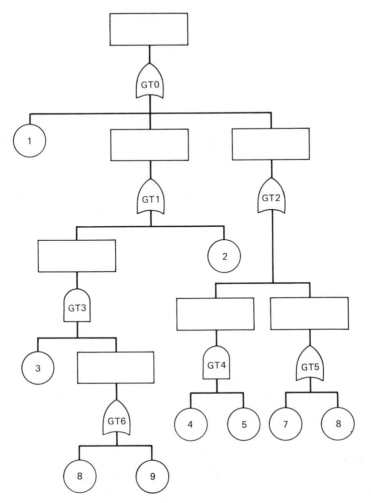

Figure 4.10 An event tree.

A simple rule to follow to develop this technique is to replace each gate by its inputs. The inputs may be the outputs of gates or basic events until all the fault tree gates are replaced with the basic event entries. At this stage the list matrix is fully completed.

For this example, to obtain a fully constructed list matrix we now replace the OR gate GT1 by its input events as separate rows, as indicated below by the dotted line. The dotted line is marked as step 2:

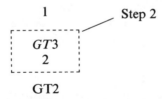

Similarly, replace, GT2, by its input events as indicated by the dotted line marked as step 3:

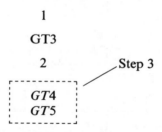

In similar fashion, we proceed with the gate, GT3. It is an AND gate, therefore, it is replaced with its input events as indicated by the following dotted line marked as step 4;

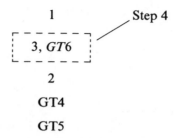

Similarly, GT4, the AND gate, is replaced by its inputs marked as step 5:

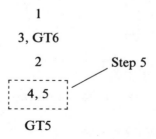

Since GT5 is an OR gate, it is replaced by its input events 7, 8 shown as step 6 below:

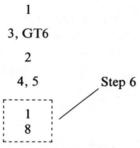

Similarly, the gate, GT6, is also an OR gate; therefore it is replaced by its input events 8 and 9 (marked as step 7) as follows:

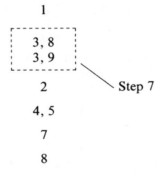

As shown above in the list matrix, the cut set 8 is a single event cut set. Therefore, eliminate cut set {3, 8} to obtain the following minimal cut sets:

1

2

7

8

3,9

4,5

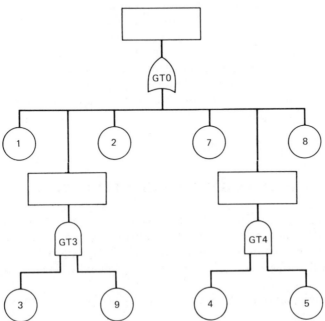

Figure 4.11 A repeated event fault tree.

Finally, if there is no repeated event in the list matrix then the cut sets generated by this method will be minimal cut sets. If this is not so, then eliminate the nonminimal cut sets (i.e., which contain other sets) from the final list matrix.

The reduced fault tree of the above list matrix is drawn in Figure 4.11. This is a repeated event free fault tree. Now one may proceed to obtain the quantitative measures of the top event.

4.7 FAULT TREE DUALITY

To reliability engineers it may be of great interest to obtain the dual fault tree. For example, in the case of top event "A system does not fail" is the dual of "system failure." Generally the occurrence of the top event is of interest more from the system safety view point to the safety analyst. The case of nonoccurrence of top event, may be of more interest to the reliability analyst.

As words "occurrence" and "nonoccurrence" of a top event suggest duality, it is simple to obtain a "success tree" from a "fault tree." To obtain a success tree (i.e., dual of a fault tree) replace all AND gates with OR gates in the original fault tree and vice-versa. In addition, the top, intermediate, and basic fault events are to be replaced by their corresponding duals (success events). In other words, the occurrence events with nonoccurrence events. For example, if the top event was "room dark" then it is to be replaced with the top event "room lit."

The minimal cut sets of the original fault tree will be minimal path sets of the dual fault tree (success tree). A path set may be defined as a set of basic events whose nonoccurrence contributes to the nonpresence of the top event. In the case of a minimal path set, it is defined as a set that cannot be further reduced and still retains its path set characteristics. The algorithm presented in the previous section to obtain minimal cut sets of a fault tree can be applied to obtain the minimal path sets of the dual fault tree.

4.8 PROBABILITY EVALUATION OF A FAULT TREE

Once the minimal cut sets or the redundancy free events of a fault tree are obtained, then one can proceed to evaluate the probability of the top event. However, before we proceed to evaluate the probability of a fault tree, we will review the basic concepts of probability laws as applied to logic gates.

Two basic operational laws of probability are presented by solving OR and AND gates.

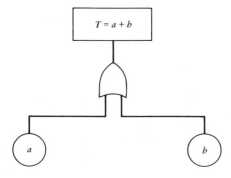

Figure 4.12 A two-input OR gate.

4.8.1 OR Gate

To explain the OR gate probability concept we are analyzing a two input OR gate as shown in Figure 4.12. For Figure 4.12, the probability expression for the top event is given by

$$P(T) = P(a) + P(b) - P(a \cdot b) \tag{4.15}$$

If a and b are statistically independent events and $P(a) \cdot P(b)$ is very small, then the above expression (4.15) can be approximated as

$$P(T) \simeq P(a) + P(b) \tag{4.16}$$

In the case of n number of inputs OR gate, the expression (4.16) may be generalized to,

$$P(a + b + c + \cdots) \simeq P(a) + P(b) + P(c) + \cdots \tag{4.17}$$

The above approximation is good if the summation of expression (4.17) is very small, which implies that the basic event probabilities $P(a)$, $P(b)$, $P(c)$, $\cdots$ are very small. However, expression (4.17) yields exact result if events $a, b, c, \cdots$ are mutually exclusive. The exact expression of (4.17) is presented in Section 4.12.

4.8.2 AND Gate

A two input events AND gate is shown in Figure 4.13. In the case of statistically independent events a and b, the multiplication rules of probability are applied to obtain the following top event probability expression:

$$P(ab) = P(a) \cdot P(b) \tag{4.18}$$

For n input AND gate, the above equation can be generalized as

$$P(a \cdot b \cdot c \cdots) = P(a) \cdot P(b) \cdot P(c) \cdots \tag{4.19}$$

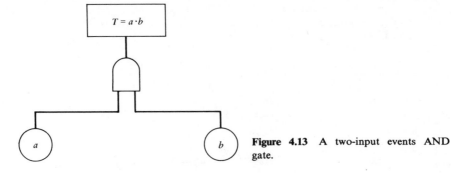

Figure 4.13 A two-input events AND gate.

Example 4. Evaluate the top event failure probability of the fault tree shown in Figure 4.14. Assume, the basic events A, B, C, D, and E are statistically independent and $P(A) = P(B) = P(C) = P(D) = P(E) = \frac{1}{4}$. The fault tree of Figure 4.14 shows that it does not have any repeated basic events. Therefore, the probability of occurrence can be evaluated at the output of each gate. However, if the repeated events in each fault tree were present then first of all one must eliminate the repeated events (i.e., obtain

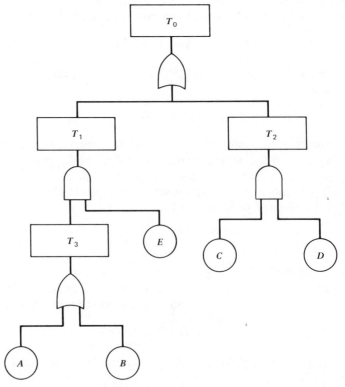

Figure 4.14 A hypothetical event tree.

minimal cut sets of the fault tree), before taking the probability of occurrence at the output of each gate.

The fault tree shown in Figure 4.14 can be solved by the following two different methods.

Method 1. Write the expression for the top event in terms of basic events. Obtain the probability of occurrence of this expression as follows. The top event expression is given by

$$T_0 = T_1 + T_2 \tag{4.20}$$

where

$$T_2 = CD \tag{4.21}$$

$$T_1 = T_3 \cdot E \tag{4.22}$$

$$T_3 = A + B \tag{4.23}$$

Hence,

$$T_0 = E(A + B) + CD \tag{4.24}$$

Therefore

$$P(T_0) = P(EA + EB + CD) \tag{4.25}$$

Now expression (4.25) can be expanded to obtain top event probability expression. If we assume the statistical occurrence of failure events then we can obtain the quantitative probability result of the top event.

Method 2. This is an alternative method to obtain the quantitative value of the top event probability by calculating the intermediate events probabilities and then using these results to obtain the top event probability result. One must note here that we assume that the failure events are statistically independent. By using expressions (4.15) and (4.18), the intermediate and top event quantitative results and expressions are as follows:

$$P(T_3) = P(A) + P(B) - P(A) \cdot P(B) = 1/4 + 1/4 - 1/16 = 7/16 \tag{4.26}$$

$$P(T_2) = P(C) \cdot P(D) = 1/4 \cdot 1/4 = 1/16 \tag{4.27}$$

$$P(T_1) = P(T_3) \cdot P(E) = 7/16 \cdot 1/4 = 7/64 \tag{4.28}$$

$$P(T_0) = P(T_1) + P(T_2) - P(T_1) \cdot P(T_2)$$

$$= 7/64 + 1/16 - 7/64 \cdot 1/16 = 169/1024 \tag{4.29}$$

$\therefore$ Probability of occurrence of top event $= 169/1024$

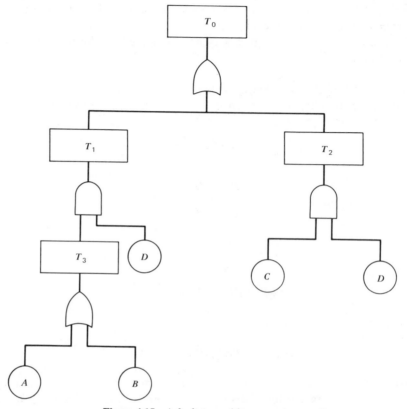

Figure 4.15 A fault tree with repeated event D.

Example 5. Suppose in Figure 4.14, the event E is replaced by event D as shown in Figure 4.15. To obtain the top event probability of the fault tree shown in Figure 4.15, we apply method 1 of the previous example. The top event expression in terms of basic events (without eliminating the repeated event D) is given by

$$T_0 = (A + B)D + CD \tag{4.30}$$

Thus,

$$T_0 = DA + BD + CD \tag{4.31}$$

By taking the probability of the top event, we get

$$P(DA + BD + CD) = P(DA) + P(BD) + P(CD) - P(DABD)$$
$$- P(DACD) - P(BDCD) + P(DABDCD) \tag{4.32}$$

The redundancy-free expression with statistically independent events is given by

$$P(DA+BD+CD)=P(A)P(D)+P(B)P(D)+P(C)P(D)$$
$$-P(D)P(A)P(B)-P(A)P(C)P(D)$$
$$-P(B)P(C)P(D)+P(A)P(B)P(C)P(D)$$

$$(4.33)$$

$$\therefore P(DA+BD+CD)=1/16+1/16+1/16-1/64$$
$$-1/64-1/64+1/256=37/256$$

The probability of occurrence of the top event is

$$37/256$$

However, if one eliminates the repeated events first then the fault tree shown in Figure 4.15 reduces to the one shown in Figure 4.16. The top event expression for Figure 4.16 becomes

$$T_0=DT_1 \qquad\qquad (4.34)$$

where

$$T_1=A+B+C \qquad\qquad (4.35)$$

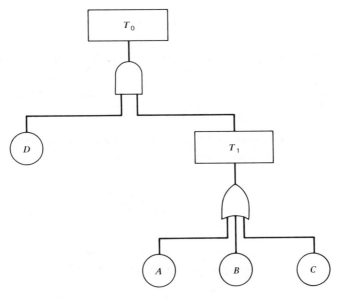

Figure 4.16   A repeated event free fault tree.

For the statistically independent events (4.34) and (4.35), probability expressions are given by

$$P(DT_1) = P(D) \cdot P(T_1) = 37/64 \cdot 1/4 = 37/256 \qquad (4.36)$$

where $P(A + B + C) = P(A) + P(B) + P(C) - P(A)P(B) - P(A)P(C) - P(B)P(C) + P(A)P(B)P(C) = 37/64$

4.8.3 Concluding Remarks

In cases where the basic event failure probabilities are very small, the inability to remove dependencies will not introduce a significant error in the end result [65]. However, one must try to remove all the dependencies in a fault tree before obtaining the final probability result.

4.9 FAILURE RATE EVALUATION OF FAULT TREES

This section outlines, how to obtain the failure rate of the fault tree top as well as the intermediate events. The following assumptions are made to develop this procedure:

1. The basic events (system components) are not repaired.
2. The fault event occurrence times (or component failure times) are exponentially distributed.
3. The fault tree is redundancy free. In other words, it contains no repeated events.
4. The basic fault occurrence or component failures are statistically independent.

The fault tree OR and AND gate failure rate expressions are developed by using the following relationship:

$$\lambda(t) = -\frac{1}{R(t)} \cdot \frac{dR(t)}{dt} \qquad (4.37)$$

where $\lambda(t) =$ the failure rate (hazard rate) at time t
$\quad\quad\quad R(t) =$ the component or system reliability

For the component constant failure rates, the OR and AND gates failure rate (hazard rate) formulas are developed in the following sections.

4.9.1 OR Gate

Logically this gate corresponds to a series system. A series system reliability can be obtained from the following equation:

$$R_S = \prod_{i=1}^{n} R_i \tag{4.38}$$

where R_i = the constant reliability of the ith component
R_S = the series system reliability
n = the number of components

When components failure times follows exponential failure laws, (4.38) becomes

$$R_S(t) = \exp - \left(\sum_{i=1}^{n} \lambda_i \right) t \tag{4.39}$$

where λ_i = constant failure rate of the ith component
t = the time

Substituting (4.39) into (4.37) yields the series system hazard rate

$$\lambda_s(t) = \sum_{i=1}^{n} \lambda_i \tag{4.40}$$

It can be recognized from the series system failure rate equation (4.40) that an OR gate output is simply the sum of its inputs.

4.9.2 AND Gate

The AND gate corresponds to a logically connected parallel configuration system. A parallel network reliability is given by the following equation:

$$R_p = 1 - \prod_{i=1}^{n} (1 - R_i) \tag{4.41}$$

where R_p = the parallel system reliability
n = the number of components
R_i = the ith component reliability

In the case of components' constant failure rates, the above equation becomes

$$R_p(t) = 1 - \prod_{i=1}^{n} (1 - e^{-\lambda_i t}) \tag{4.42}$$

where λ_i = the ith component constant failure rate
     t = the time

After substituting (4.42) into (4.37), we get the following [53] results:

$$\lambda_p(t) = \left\{ \sum_{j=1}^{n} \lambda_j(z_j - 1) \right\} \left\{ \prod_{j=1}^{n} z_j - 1 \right\}^{-1} \qquad (4.43)$$

where $1/z_j = (1 - e^{-\lambda_j t})$ for $j = 1, 2, 3, \ldots, n$.

Example 6.   Evaluate top event failure rate of the fault tree shown in Figure 4.17, for a 100-hour mission. Assume

$$\lambda_1 = \lambda_2 = \lambda_3 = \lambda_4 = \lambda_5 = \lambda_6 = \lambda_7 = 0.001 \text{ failure/hour}$$

By utilizing (4.40), the output event failure rates of OR gates GT1, GT3,

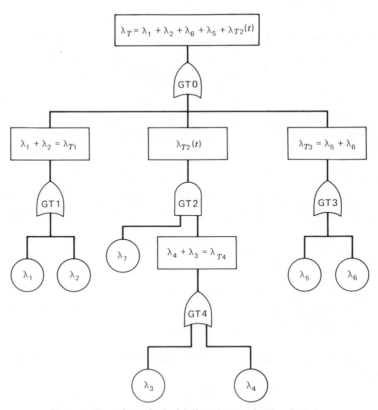

Figure 4.17   A hypothetical failure rate evaluation fault tree.

GT4, and GT0 are evaluated as follows:

$$\lambda_{T0} = \lambda_{T1} + \lambda_{T2}(t) + \lambda_{T3} = 0.0040082 \text{ failure/hour} \tag{4.44}$$

$$\lambda_{T1} = \lambda_1 + \lambda_2 = 0.002 \text{ failure/hour} \tag{4.45}$$

$$\lambda_{T3} = \lambda_5 + \lambda_6 = 0.002 \text{ failure/hour} \tag{4.46}$$

and

$$\lambda_{T4} = \lambda_3 + \lambda_4 = 0.002 \text{ failure/hour} \tag{4.47}$$

Similarly, we utilize (4.43) to obtain the output event failure rate of the AND gate GT2 as follows for a 100-hour mission:

$$\lambda_{T2}(t) = \frac{\lambda_7(z_7 - 1) + (z_{T4} - 1)\lambda_{T4}}{z_7 z_{T4} - 1} = 0.0000082 \text{ failure/hour} \tag{4.48}$$

where $z_i = 1/(1 - e^{-\lambda_i t})$ for $i = 7, T4$.

When an AND gate output event is an input event to another AND gate then the hazard rates of all the intermediate (including the top event) events can only be obtained from the reliability function of these events. In other words, the hazard rate or failure rate result obtained for an AND gate output event cannot be used as an input to another AND gate.

If two or more AND gates are encountered in series, it is strongly advised to establish the reliability function at the output event level of each gate then apply the hazard rate formula of (4.43).

Example 7. A two-AND-gates-in-series fault tree, shown in Figure 4.18, is required to compute the failure rate of the top event for a 100-hour mission. Assume $\lambda_1 = \lambda_2 = \lambda_3 = 0.001$ failure/hour and the basic failures are statistically independent. By utilizing (4.43) we get

$$\lambda_{GT1}(t) = \frac{2\lambda}{z+1} = 0.00018 \text{ failure/hour} \tag{4.49}$$

where $z = 1/(1 - e^{-\lambda t})$.

Gates GT1 and GT0 output event unreliability and reliability equations are given by

$$P_{GT1}(t) = P_1(t) \cdot P_2(t) \tag{4.50}$$

$$R_{GT0}(t) = 1 - P_1(t) \cdot P_2(t) \cdot P_3(t) \tag{4.51}$$

where $P_i(t)$ = the unreliability of the event i at time t for $i = 1, 2, 3$
$P_{GT1}(t)$ = the unreliability of the gate, GT1 output event
$R_{GT0}(t)$ = the reliability of the top event

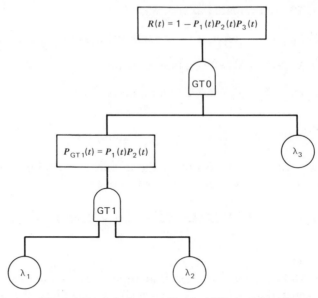

Figure 4.18   A fault tree with two AND gates in series.

To obtain the top event hazard rate, substitute (4.51) into (4.37). Since $\lambda_1 = \lambda_2 = \lambda_3$, we use (4.43) to obtain the following top event hazard rate result:

$$\lambda_{GT0}(t) = \frac{3\lambda}{z^2 + z + 1} = 0.00002 \text{ failure/hour} \tag{4.52}$$

where $z = 1/(1 - e^{-\lambda t})$.

4.10   FAULT TREE EVALUATION OF REPAIRABLE COMPONENTS

In this section we are concerned only with the fault tree evaluation of repairable components [22]. This type of situation is frequently encountered in real life where the system components are normally repaired whenever they fail. The method presented in this paper assumes that the component failures are statistically independent and that the component failure and repair rates are constant; in addition, the repaired system components are considered as good as new. Furthermore, this method is only applicable to the cases where one may be concerned with calculating the top or intermediate event steady-state unavailability, limiting mean repair rate, limiting mean failure rate, and steady-state failure rate. Another major assumption of this method is that it assumes that fault trees are redundancy free, (i.e., no basic repeated events are allowed).

In most cases a redundancy-free expression can be obtained by applying basic Boolean reduction techniques. For certain cases it is simpler to obtain Boolean indicated cut sets (BICS) [33] and then eliminate the redundant cut sets by inspection. It may also be useful to eliminate as many repeated (redundant) events as possible at the fault tree construction stage and eliminate the remaining ones with the Boolean reduction techniques. However, if some of the repeated events are impossible to eliminate and if the probability of occurrence for basic events is less than 0.1 [65], the error generated in the end result will be either negligible or of very small magnitude.

The main advantage of applying this technique is that, the original dependency free fault tree is unchanged; and the OR and AND gate steady-state unavailability, limiting mean failure rate, limiting mean repair rate, and limiting steady-state failure rate formulas, can be applied directly to both the intermediate and top events of the fault tree. These formulas [128] for the OR and AND gates are discussed in the following sections.

4.10.1 OR Gate

This gate simply represents a series system with n nonidentical repairable components. The OR gate output event unavailability $\overline{A}_s$, can be obtained from the following equation:

$$\overline{A}_s = 1 - \prod_{i \in X} \left(1 - \overline{A}_i\right) \tag{4.53}$$

where $\overline{A}_i$ = the unavailability of the repairable component i
 X = a set of n number of components

For a repairable component with constant failure and repair rates the equation for the unavailability $\overline{A}$ may be expressed [129] as follows:

$$\overline{A}(t) = \frac{\lambda}{\mu + \lambda} \left(1 - e^{-(\lambda + \mu)t}\right) \tag{4.54}$$

where t = time
 λ = the component failure rate
 μ = the component repair rate

For large t, the above equation becomes

$$\overline{A} = \frac{\lambda}{\lambda + \mu} \tag{4.55}$$

By substituting (4.55) into (4.53) we obtain for the series system

$$\overline{A}_s = 1 - \prod_{i \in X} \frac{\mu_i}{\lambda_i + \mu_i} \tag{4.56}$$

Similarly, the following OR gate output event, limiting failure rate, limiting mean failure rate, and limiting mean repair rate equations, is

$$\lambda_{ss} = \{\text{series system steady state availability}, (1 - \bar{A}_s)\}$$

$$\times \{\text{series system failure rate}, \hat{\lambda}_{sm}\}$$

$$= (1 - \bar{A}_s) \sum_{i \in X} \lambda_i \tag{4.57}$$

where λ_{ss} is the series system steady-state failure frequency:

$$\hat{\lambda}_{sm} = \sum_{i \in X} \lambda_i \tag{4.58}$$

where $\hat{\lambda}_{sm} =$ the series system limiting mean failure rate
$\hat{\mu}_{sm} = \{\text{series system steady-state availability}, (1 - A_s)\} \times (\text{series system failure rate}, \hat{\lambda}_{sm}) / (\text{series system unavailability}, \bar{A}_s)$.

$$\hat{\mu}_{sm} = \frac{\lambda_{ss}}{\bar{A}_s} \tag{4.59}$$

where $\hat{\mu}_{sm}$ is the series system limiting mean repair rate.

4.10.2 AND Gate

An AND gate is the representation of a parallel system composed of n (number) of nonidentical components. Since the parallel system is a dual of the series system AND gate output event, steady-state unavailability, steady-state failure rate, limiting mean failure rate, and limiting mean repair rate equations can be obtained directly from (4.56), (4.57), (4.58), and (4.59):

$$\bar{A}_p = \prod_{i \in X} \left\{ 1 - \frac{\mu_i}{\mu_i + \lambda_i} \right\} \tag{4.60}$$

where p denotes a parallel system.

$$\lambda_{ps} = \sum_{i \in X} \mu_i (\bar{A}_p) \tag{4.61}$$

$$\hat{\lambda}_{pm} = \frac{\lambda_{ps}}{1 - \bar{A}_p} \tag{4.62}$$

and

$$\hat{\mu}_{pm} = \sum_{i \in X} \mu_i \tag{4.63}$$

Similarly, the respective equations, in the case of a *m*-out-of-*n* identical inputs AND gate are

$$\bar{A}_{m/n} = \sum_{i=m}^{n} \binom{n}{i} (\bar{A})^i (1-\bar{A})^{n-i} \tag{4.64}$$

$$\lambda_{m/n} = \frac{n!}{(n-m)!(m-1)!} \frac{(1/\lambda)^{m-1}}{(1/\mu)^m} \bar{A}^n \tag{4.65}$$

$$\hat{\lambda}_{m/n} = \frac{m!(1/\lambda)^{m-1}(1/\mu)^{n-m}}{(n-m)!(m-1)! \sum_{i=m}^{n} \binom{n}{i}(1/\lambda)^i(1/\mu)^{n-i}} \tag{4.66}$$

and

$$\hat{\mu}_{m/n} = \frac{n!(1/\lambda)^{m-1}(1/\mu)^{n-m}}{(n-m)!(m-1)! \sum_{i=0}^{m-1} \binom{n}{i}(1/\lambda)^i(1/\mu)^{n-i}} \tag{4.67}$$

It is easily seen from the above equations that, for identical inputs OR and AND gate, the equations are special cases of the *m*-out-of-*n* inputs AND gate equations. In the case of an AND gate, *m* takes on the value of the number of inputs to that AND gate, whereas in the case of an OR gate, *m* is equal to unity.

Example 8. Suppose the objective in Figure 4.19 is to obtain the top event steady-state unavailability, steady-state failure frequency, limiting mean failure rate, and limiting mean repair rate. Assume that all of the basic events of the fault tree have the same failure and repair rate respectively, that is, $\lambda = 0.001$ failure/hour; and $\mu = 0.05$ repair/hour. Furthermore, assume that all of the basic events are statistically independent.

From (4.55) the single component steady-state unavailability is

$$\bar{A} = \frac{\lambda}{\lambda + \mu} = \frac{0.001}{0.051} \cong 0.02$$

In the case of an OR gate output event GT1, the unavailability from (4.53) is

$$\bar{A}_s \cong 0.04$$

From (4.57),

$$\lambda_{ss} = (\lambda_1 + \lambda_2) \prod_{i=1}^{2} \frac{\mu_i}{\mu_i + \lambda_i}$$

$$= 0.0019 \text{ failure/hour}$$

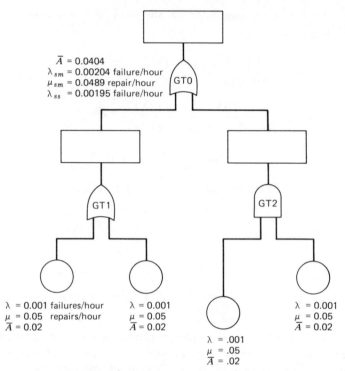

Figure 4.19 A hypothetical fault tree.

Equation 4.58 yields

$$\hat{\lambda}_{sm} = 0.002 \text{ failure/hour}$$

The limiting mean repair rate $\hat{\mu}_{sm}$ is obtained from (4.59)

$$\hat{\mu}_{sm} = \frac{\left(\sum_{i=1}^{2} \lambda_i \right) \left(\prod_{i=1}^{2} \frac{\mu_i}{\mu_i + \lambda_i} \right)}{1 - \prod_{i=1}^{2} \mu_i / (\lambda_i + \mu_i)}$$

$$= 0.0475 \text{ repair/hour}$$

Similarly, for the AND gate output event, GT2, and top event OR gate GT0, the following information was calculated from (4.60), (4.61), (4.62), (4.63), (4.56), (4.57), (4.58), and (4.59), respectively. In the case of an AND gate GT2,

$$\bar{A}_p = 0.0004$$

$$\lambda_{ps} = 0.00004 \text{ failure/hour}$$

$$\hat{\lambda}_{pm} = 0.000041 \text{ failure/hour}$$

and

$$\hat{\mu}_{pm} = 0.01 \text{ repair/hour}$$

Similarly, in the case of the top event OR gate, GT0

$$\bar{A}_s = 0.0404$$

$$\lambda_{ss} = 0.00195 \text{ failure/hour}$$

$$\hat{\lambda}_{sm} = 0.00204 \text{ failure/hour}$$

and

$$\hat{\mu}_{sm} = 0.0489 \text{ repair/hour}$$

4.11 LAMBDA TAU METHOD

This is another method that takes into consideration the repair of the basic components. The Lambda Tau technique requires redundant-free expressions from the fault tree diagram. In other words the basic events of the tree must not be repeated events. In many cases it may be obtained by Boolean substitution reduction techniques. However, this method incorporates many other restrictions. The Lambda Tau method calculations for an AND gate are based on the coexistence of all failures, and the calculations for an OR gate are based upon at least one failure among n number of possible failures. The basic formulas for the AND and OR gate parameters are derived in flow research references 124 and 65. The main restrictions of this technique are (a) τ/T is small, where τ is repair time of a component in question, where T is the time interval of interest; (b) the basic event failure rates are very small; (c) the product of the failure rate and repair time is very small (i.e., must be less than 1); (d) the product of the failure rate and the mission time is very small (i.e., must be less than 1 preferably 0.1); (e) the failures and repair rates are constant; and (f) failures occur independently.

The basic formulas for reliability of the AND (AND Priority) OR gates are derived in reference 126. AND and OR gate parameter formulas are presented in the following sections.

4.11.1   AND Gate (Coexistence of All Failures)

The general formula of the probability, P_{AND}, that n failures coexist in a small time interval, dt for the first time can be obtained:

$$P_{AND} = \prod_{i=1}^{n} \lambda_i \left[\prod_{i=2}^{n} \tau_n + \tau_1 \tau_3 \cdots \tau_n + \cdots \tau_1 \tau_2 \cdots \tau_{n-1} \right] t \qquad (4.68)$$

where n is the number of components and λ_i is the constant failure rate of the ith component.

The AND gate output event hazard rate (failure rate) and repair time equations are given by

$$\lambda_{AND} = \prod_{i=1}^{n} \lambda_i \left[\prod_{i=2}^{n} \tau_i + \prod_{\substack{i=1 \\ i \neq \text{even}}}^{n} \tau_i + \cdots \prod_{i=1}^{n-1} \tau_i \right] \qquad (4.69)$$

and

$$\tau_{AND} = \frac{1}{\sum\limits_{i=1}^{n} \tau_i} \qquad (4.70)$$

It is emphasized that (4.68), (4.69), and (4.70) are only valid for assumptions outlined in the earlier section

4.11.2   OR Gate (At Least One Failure Among n Possible Failures)

This gate represents a system with n components connected in a series configuration. The probability that one or more failures occur is

$$P_{OR}(t) = 1 - e^{(\sum_{i=1}^{n} \lambda_i)t} \qquad (4.71)$$

The OR gate output event failure rate and repair time are

$$\lambda_{OR} = \sum_{i=1}^{n} \lambda_i \qquad (4.72)$$

and

$$\tau_{OR} = \frac{\sum\limits_{i=1}^{n} \lambda_i \tau_i}{\sum\limits_{i=1}^{n} \lambda_i} \qquad (4.73)$$

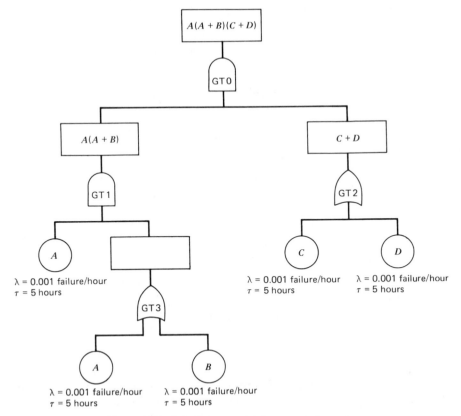

Figure 4.20 A fault tree containing repeated events.

As for the AND gate output event formulas, these equations are only valid under the assumptions outlined in the sections earlier.

Example 9. A fault tree containing a repeated event is shown in Figure 4.20. Assume the occurrence of basic fault events is statistically independent; then obtain the top event quantitative measures of the Lambda Tau technique.

As it can be realized from the fault tree that the repeated event A has to be eliminated before we can apply the Lambda Tau technique to compute quantitative reliability measures.

The output event expression of the gate, $GT1$, can be simplified by applying the following Boolean identity:

$$A(A+B)=A \tag{4.74}$$

Therefore the simplified fault tree of Figure 4.20 becomes as shown in Figure 4.21. To determine the top event quantitative measures of the

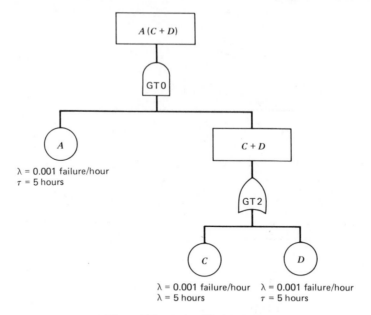

Figure 4.21   A simplified fault tree.

reduced fault tree shown in Figure 4.21, the OR gate, *GT*2, output event failure rate and repair time are obtained by using (4.72) and (4.73)

$$\lambda_{GT2}=2\lambda=0.002 \text{ failure/hour} \tag{4.75}$$

$$\tau_{GT2}=\tau=5 \text{ hours} \tag{4.76}$$

To obtain the quantitative measures of the top event, failure rate and repair time, use expressions (4.69) and (4.70), respectively:

$$\lambda_{GT0}=\lambda_A\lambda_{GT2}(\tau_A+\tau_{GT2})=0.00002 \text{ failure/hour} \tag{4.77}$$

$$\tau_{GT0}=\frac{\tau_{GT2}\tau_A}{\tau_{GT2}+\tau_A}=2.5 \text{ hours} \tag{4.78}$$

For a 100-hour mission, the top event probability that *n* failures coexist in time interval *dt* for the first time is

$$P_{GT0}=\lambda_{GT0}t$$

$$=0.00002\times100=0.002/\text{mission} \tag{4.79}$$

where λ_{GT0} is obtained from (4.77).

4.12 REPAIRABLE COMPONENT FAULT TREE EVALUATION WITH KINETIC THEORY

This is another method to evaluate reliability indices of the fault trees with repairable components. Before applying the kinetic theory, the minimal cut sets of a fault tree are to be determined. This approach was originated in reference 68. In this section assume that the component failures are statistically independent. The major steps to be followed for this technique are outlined below:

Step 1. Construct the fault tree of a device, a subsystem, or a system in question.

Step 2. Determine minimal cut sets of the constructed fault tree.

Step 3. Develop each primary event information of a minimal cut set fault tree.

Step 4. Similarly, develop each cut set information of a minimal cut set fault tree in question.

Step 5. Finally, evaluate the fault tree top event information.

To obtain basic cut set and top events quantitative reliability information the following notations are used.

Basic Events.

$\lambda =$ the constant failure rate of the basic event or component
$\mu =$ the constant repair rate of the basic event or component
$t =$ mission time
$F(t) =$ probability of a component failed condition at time t
$F_f(t) =$ probability that a component has its first failure by time t
$W =$ probability that a component fails or a basic fault event occurs in time interval $[t, t+\Delta t]$
$W_f =$ probability that a component has its first failure in time interval $[t, t+\Delta t]$

Cut-Sets.

$\lambda'(t) =$ the cut set failure rate at time t
$\mu'(t) =$ the cut set repair rate at time t

Top Event.

$\lambda_T(t) =$ the top event failure rate at time t
$\mu_T(t) =$ the top event repair rate at time t

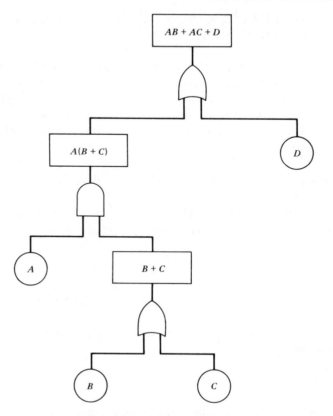

Figure 4.22   A hypothetical fault tree.

In a detailed form the kinetic tree theory is described in reference 69. Here we simply deal with the practical aspect of this theory, with assumptions that the basic failure rate, and repair rates are constant and the failures are statistically independent [130]. To demonstrate the practicality of this approach, the following hypothetical example is presented in Figure 4.22.

Example 10.   For the fault tree shown in Figure 4.22, it is necessary to obtain the top event unavailability and failure and repair rate information. For the fault tree shown we develop the required information [130] for the basic events, cut-sets, and top event. One should note here that the constructed fault tree has no repeated events.

BASIC FAILURE EVENT INFORMATION.   Assume that a repairable component has constant failure and repair rates; therefore, by applying the Markov process concept we obtain the following differential-difference equations

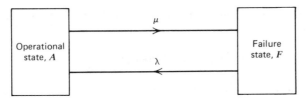

Figure 4.23   A single component state space diagram.

for the component operational and failure states as shown in Figure 4.23.

$$A(t+\Delta t)=(1-\lambda \Delta t)A(t)+\mu \Delta t \, F(t) \qquad (4.80)$$

$$F(t+\Delta t)=(1-\mu \Delta t)F(t)+\lambda \Delta t \, A(t) \qquad (4.81)$$

In the limiting case the above equations become

$$\frac{dA(t)}{dt}=-\lambda A(t)+\mu F(t) \qquad (4.82)$$

$$\frac{dF(t)}{dt}=-\mu F(t)+\lambda A(t) \qquad (4.83)$$

At $A(0)=1$, other initial condition probabilities are equal to zero, where $A(t)$ is the component availability at time t and $F(t)$ is the component unavailability at time t. By solving the above differential equations we get

$$A(t)=\frac{\mu}{\lambda+\mu}+\frac{\lambda}{\lambda+\mu}e^{-(\lambda+\mu)t} \qquad (4.84)$$

$$F(t)=\frac{\lambda}{\lambda+\mu}-\frac{\lambda}{\lambda+\mu}e^{-(\lambda+\mu)t} \qquad (4.85)$$

For large t, (4.85) becomes

$$F=\frac{\lambda}{\lambda+\mu} \qquad (4.86)$$

To obtain the failure probability at time t, set $\mu=0$ in (4.85):

$$F_f(t)=1-e^{-\lambda t} \qquad (4.87)$$

For small λt the above equation may be approximated to

$$F_f(t)\simeq\lambda t \qquad (4.88)$$

Now, we define the following expression for the probability that a component fails in a time interval $[t, t+\Delta t]$:

$$W(t)\Delta t=[1-F(t)]\lambda\,\Delta t \qquad (4.89)$$

One should note that (4.89) is the second part of the right-hand side of (4.81). By substituting (4.85) into (4.89) for large t we get

$$W=W(t)\Delta t=\frac{\mu}{\lambda+\mu}(\lambda\,\Delta t) \qquad (4.90)$$

Similarly, for a nonrepairable component (i.e., $\mu=0$) substitute (4.87) into (4.89) for small λt; we get

$$W_f=W_f(t)\cdot\Delta t=\lambda\,\Delta t \qquad (4.91)$$

CUT SET INFORMATION.   To demonstrate how to obtain the cut set information we will use the fault tree example shown in Figure 4.22. The top event expression of Figure 4.22 is composed of the following cut sets:

$$\text{Top event}=AB+AC+D \qquad (4.92)$$

Now, consider the cut set AB in (4.92). Here, we are interested to find the probability of first failure in time interval $[t, t+\Delta t]$.

There are two possibilities to encounter failure of cut set, AB, in a small interval Δt (here we assume that only one failure occurs in a small time Δt):

1. A is in failed state and B fails in Δt
2. B is in failed state and A fails in Δt

Thus we define, W_{AB}, as the probability of first failure of the cut set AB in the time interval $[t, t+\Delta t]$. Therefore,

$$W_{AB}=F_A W_B+F_B W_A \qquad (4.93)$$

In the case of repairable events A and B substitute (4.86) and (4.90) into (4.93):

$$W_{AB}=\frac{\lambda_A}{(\lambda_A+\mu_A)}\cdot\frac{\lambda_B\mu_B}{(\lambda_B+\mu_B)}\cdot\Delta t$$

$$+\frac{\lambda_B}{(\lambda_B+\mu_B)}\cdot\frac{\lambda_A}{(\lambda_A+\mu_A)}\mu_A\cdot\Delta t \qquad (4.94)$$

$$=\frac{\lambda_A\lambda_B}{(\lambda_A+\mu_A)(\lambda_B+\mu_B)}\{\mu_B+\mu_A\}\Delta t \qquad (4.95)$$

Similarly, for cut sets AC and D we obtain the following expressions:

$$W_{AC} = F_A W_C + F_C W_A \tag{4.96}$$

$$W_{AC} = \frac{\lambda_A \lambda_C}{(\lambda_A + \mu_A)(\lambda_C + \mu_C)} \{\mu_C + \mu_A\} \Delta t \tag{4.97}$$

and

$$W_D = W_D \tag{4.98}$$

$$W_D = \frac{\lambda_D \mu_D}{(\lambda_D + \mu_D)} \cdot \Delta t \tag{4.99}$$

To determine probability that the cut sets AB, AC, and D are in failed state one should multiply the individual event probabilities for the statistically independent events. Thus, by utilizing (4.86), we obtain

$$F_{AB} = F_A F_B = \frac{\lambda_A \lambda_B}{(\lambda_A + \mu_A)(\lambda_B + \mu_B)} \tag{4.100}$$

$$F_{AC} = F_A F_C = \frac{\lambda_A \lambda_C}{(\lambda_A + \mu_A)(\lambda_C + \mu_C)} \tag{4.101}$$

$$F_D = \frac{\lambda_D}{\lambda_D + \mu_D} \tag{4.102}$$

Suppose, if event B of the cut set AB is not repaired, then we will denote with a small alphabetic letter, b. Therefore one may rewrite (4.93) as

$$W_{Ab} = F_A W_b + F_b W_A \tag{4.103}$$

In the case of repairable and nonrepairable events A and b, respectively, substitute equations (4.91), (4.90), (4.86), and (4.88) into (4.103) to obtain

$$W_{Ab} = \frac{\lambda_A}{\lambda_A + \mu_A} \cdot (\lambda_b \Delta t) + (\lambda_b t) \frac{\mu_A \lambda_A}{\mu_A + \lambda_A} \Delta t \tag{4.104}$$

Hence,

$$W_{Ab} = \frac{\lambda_A \lambda_b}{\lambda_A + \mu_A} (1 + t\mu_A) \Delta t \tag{4.105}$$

To obtain cut set failure rate we use (4.89)

$$\lambda'(t) = \frac{W(t)}{1 - F(t)} \tag{4.106}$$

Similarly, the cut set repair rate is obtained by using equation (4.107):

$$\mu'(t) = \frac{W(t)}{F(t)} \tag{4.107}$$

Since

$$W = W(t)\,dt \tag{4.108}$$

Now consider the cut set AB, since W_{AB} is known from (4.95) therefore, we substitute (4.95) into (4.108) to get

$$W_{AB}(t) = \frac{\lambda_A \lambda_B}{(\lambda_A + \mu_A)(\lambda_B + \mu_B)}\{\mu_A + \mu_B\} \tag{4.109}$$

To obtain cut set AB repair rate, substitute (4.100) and (4.109) into (4.107) to get

$$\mu'_{AB} = \mu_A + \mu_B \tag{4.110}$$

Similarly, to obtain cut set failure rate, substitute (4.100) and (4.109) into (4.106) to get

$$\lambda'_{AB} = \frac{\lambda_A \lambda_B (\mu_A + \mu_B)}{\lambda_A \mu_B + \mu_A \lambda_B + \mu_A \mu_B} \tag{4.111}$$

In similar fashion, one can obtain the failure and repair rates for cut sets AC and D as follows:

$$\lambda'_{AC} = \frac{\lambda_A \lambda_C (\mu_A + \mu_C)}{\lambda_A \mu_C + \mu_A \lambda_C + \mu_A \mu_C} \tag{4.112}$$

$$\lambda'_D = \lambda_D \tag{4.113}$$

and

$$\mu'_{AC} = (\mu_A + \mu_C) \tag{4.114}$$

$$\mu'_D = \mu_D \tag{4.115}$$

TOP EVENT INFORMATION. To obtain the top event probability information, one has to take advantage of the union of the minimal cut sets, since the occurrence of any one of the cut sets will cause the top event to occur.

The probability of the union of the top events is given by

$$P(T_1+T_2+\cdots+T_n)=\left[\,P(T_1)+P(T_2)+\cdots+P(T_n)\,\right] \qquad \leftarrow n \text{ terms}$$

$$-\left[\,P(T_1T_2)+P(T_1T_3)+\cdots+P\!\left(\,\underset{i\neq j}{T_i\,T_j}\,\right)\right] \qquad \leftarrow\binom{n}{2}\text{ terms}$$

$$+\left[\,P(T_1T_2T_3)+P(T_1T_2T_4)+\cdots+P\!\left(\,\underset{i\neq j\neq k}{T_iT_jT_k}\,\right)\right] \qquad \binom{n}{3}\text{ terms}$$

$$\cdots(1)^{n-1}\left[\,P(T_1T_2\cdots T_n)\,\right] \qquad \leftarrow\binom{n}{n}\text{ term} \qquad (4.116)$$

Consider now the following top event minimal cut set expression of Figure 4.22:

$$T=AB+AC+D \qquad (4.117)$$

The probability expression of the above expression becomes

$$F(AB+AC+D)=F(AB)+F(AC)+F(D)-F(ABC)-F(ABD)$$
$$-F(ACD)+F(ABCD) \qquad (4.117)$$

For statistically independent events

$$F_{\text{TOP}}=F(AB+AC+D)=F_AF_B+F_AF_C+F_D-F_AF_BF_C-F_AF_BF_D$$
$$-F_AF_CF_D+F_AF_BF_CF_D \qquad (4.118)$$

Now consider (4.117); it contains event A, which is common to both cut sets AB and AC. The occurrence of this common event A will cause the simultaneous failure of cut sets AB and AC, if the component A fails in interval $[t, t+\Delta t]$.

The probability expression of this intersection is given by:

$$W_{ABC}=W_AF_BF_C \qquad (4.119)$$

Therefore by substituting (4.86) and (4.90) into the above expression we get

$$W_{ABC}=\frac{\lambda_A\lambda_B\lambda_C\mu_A}{(\lambda_A+\mu_A)(\lambda_B+\mu_B)(\lambda_C+\mu_C)}\cdot\Delta t \qquad (4.120)$$

When obtaining, W_{TOP}, one should be careful that it is composed of two states:

1. All cut sets are operating at time t.
2. A cut set fails in a time interval $[t, t+\Delta t]$.

Thus we write an expression for W_{TOP} as follows:

$$W_{TOP} = W_{AB}(1-F_C)(1-F_D) + W_{AC}(1-F_B)(1-F_D)$$

$$+ W_D(1-F_A F_B)(1-F_A F_C)$$

$$- W_{ABC}(1-F_D) \tag{4.121}$$

The term $W_D(1-F_A F_B)(1-F_A F_C)$ of (4.121) becomes in simplified form

$$W_D(1-F_A F_B - F_A F_C - F_A F_B F_C)$$

The top event failure and repair rates, λ_{TOP} and μ_{TOP}, for Figure 4.22 may be obtained by substituting (4.121) and (4.118) into the following expressions:

$$\lambda_{TOP} = \frac{W_{TOP}}{\Delta t(1-F_{TOP})} \tag{4.122}$$

and

$$\mu_{TOP} = \frac{W_{TOP}}{(\Delta t)F_{TOP}} \tag{4.123}$$

4.13 ADVANTAGES AND DISADVANTAGES OF THE FAULT TREE TECHNIQUE

Like any other technique, the fault tree technique has its advantages and disadvantages:

Advantages.

1. It provides insight into the system behavior.
2. It requires the reliability analyst to understand the system thoroughly and deal specifically with one particular failure at a time.
3. It helps to ferret out failures deductively.
4. It provides a visibility tool to designers, users, and management to justify design changes and trade-off studies.
5. It provides options to perform quantitative or qualitative reliability analysis.
6. This technique can handle complex systems more easily.

Disadvantages.

1. This is a costly and time-consuming technique.
2. Its results are difficult to check.
3. This technique normally considers that the system components are in either working or failed state. Therefore, the partial failure states of components are difficult to handle.
4. Analytical solutions for fault trees containing stand-bys and repairable priority gates are difficult to obtain for the general case.
5. To include all types of common-cause failures it requires a considerable effort.

4.14 COMMON-CAUSE FAILURES

As the field of reliability engineering is becoming a recognized discipline in engineering so is the awareness of associated problems such as common-cause failures, which were overlooked some years ago. In recent years the common-cause failures have received widespread attention for reliability analysis of redundant components, units or systems, because the assumption of statistical-independent failure of redundant units is easily violated in practice [93]. It may easily be verified from reference 116. This paper reports frequency of common-cause failure in the U. S. power reactor industry: "Of 379 components failures or groups of failures arising from independent causes, 78 involved common-cause failures of two or more components."

A common-cause failure is defined in reference 105 as any instance where multiple units or components fail due to a single cause. Some of the common-cause failures may occur due to:

1. *Equipment design deficiency.* This includes those failures that may have been overlooked during the design phase of the equipment or system, and may be due to the interdependence between electrical and mechanical subsystems or components of a redundant system.
2. *Operations and maintenance errors.* These errors may occur due to improper adjustment or calibration, carelessness, improper maintenance, etc.
3. *External normal environment.* This includes causes such as dust, dirt, humidity, temperature, moisture, and vibration. These may be the normal extremes of the operating environment.
4. *External catastrophe.* This includes natural external phenomena such as flood, earthquake, fire, and tornado. The occurrence of any one of these events may affect the redundant system at a plant.

5. *Common manufacturer.* The redundant equipment or component procured from the same manufacturer may have the same design or fabrication errors. For example, the fabrication errors may occur due to use of wrong material, wiring a circuit board backward, poor soldering, etc.

6. *Common external power source.* A common-cause failure may occur due to the common external power source of the redundant equipment, subsystem, or unit.

7. *Functional deficiency.* This may occur due to inappropriate instrumentation or inadequacy of designed protective action.

There are several examples of common-cause failures in nuclear power systems [114]. Some spring loaded relays in a parallel configuration fail simultaneously due to a common cause. Furthermore, due to a maintenance error of incorrectly disengaging the clutches, two motorized valves are placed in a failed state. In addition, a steam line rupture causing multiple circuit board failures is another example. The common cause is the steam line rupture in this case. In some cases instead of triggering a complete redundant system failure (simultaneous failure), which is the extreme case, the common cause may produce a less severe but *common*, degradation of the redundant unit. This will increase the joint probability of failure of the system units. It may be due to harsh accident environment. In this degradation state, the redundant unit may fail at a time later than the first unit failure. Because of the common morose environment, the second unit failure is *dependent* and *coupled* to the first unit failure.

Although the existence of common-cause failures has been recognized for a long time, no concrete steps were taken to represent them systematically until the late 1960s. Most of the literature on the subject is presented in bibliography on common-cause failures [93].

Some of the newly established theory and models to analyze common-cause are presented in this section.

4.14.1 Common-Cause Failure Analysis of Reliability Networks

In this section we present a newly developed method [88, 101] to analyze active identical units with statistically independent and dependent (common-cause) failures. However, this method may be extended to other reliability models and probability densities. To develop this method, it was assumed that each unit has a certain amount of common-cause failures.

Since from past experience [101] it is known that the common-cause failures occur in real life, the parameter α is introduced into the newly developed formulas to include common-cause failures [88]. The parameter α can be obtained from the operating experience data of the redundant system or equipment

$$\alpha \equiv \text{fraction of unit failures that are common cause}$$

The above parameter can be considered a point estimate of the conditional probability that a unit failure is common cause. A unit failure rate λ can be considered to have two mutually exclusive components, λ_1 and λ_2, that is,

$$\lambda = \lambda_1 + \lambda_2 \tag{4.124}$$

where λ_1 = the unit independent mode constant failure rate.
        λ_2 = the redundant system or unit constant common-cause failure rate

Since

$$\alpha = \frac{\lambda_2}{\lambda} \tag{4.125}$$

$$\therefore \lambda_2 = \alpha\lambda \tag{4.126}$$

and λ_1 can be obtained from (4.124) by substituting (4.126)

$$\therefore \lambda_1 = (1-\alpha)\lambda \tag{4.127}$$

The system reliability, hazard rate, and MTTF formulas as well as the graphical plots are developed for a parallel, k-out-of-n, series, and a bridge network as discussed in the following sections.

A Parallel Network. The modified identical units parallel network is shown in Figure 4.24. It is simply a parallel network with a unit in series. The parallel stage (i.e., labeled "1") of Figure 4.24 represents all the independent failures for any n unit system. The series unit stage labeled "2" in Figure 4.24 represents all the common-cause failures of the system.

The common-cause failure probability hypothetical unit is connected in series with the independent failure mode units. A failure of the hypothetical series unit (i.e., the common-cause failure) will cause the system failure. It is assumed that all the common-cause failures are completely coupled. The system reliability R_p of the Figure 4.24 can be written as

$$R_p = \left\{ 1 - (1-R_1)^n \right\} R_2 \tag{4.128}$$

where n = the number of identical units
        R_1 = the unit's independent failure mode reliability
        R_2 = the system common failure mode reliability

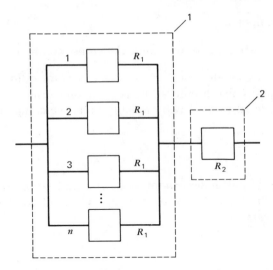

Figure 4.24 A modified identical units parallel network.

For constant failure rates λ_1 and λ_2 from (4.126) and (4.127) and for reliabilities R_1 and R_2, the (4.128) can be rewritten as

$$R_p(t) = \left\{ 1 - (1 - e^{-(1-\alpha)\lambda t})^n \right\} e^{-\alpha \lambda t} \tag{4.129}$$

where t is the time.

The reliability plots of (4.129) are shown for $n = 2, 3, 4$, in Figures 4.25, 4.26, and 4.27, respectively. These plots clearly show the effect of common-

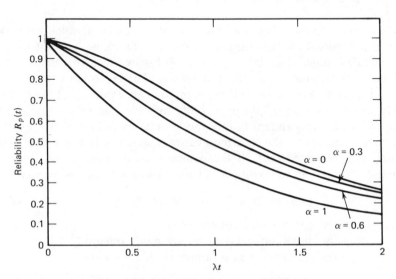

Figure 4.25 A two-parallel-units reliability plot.

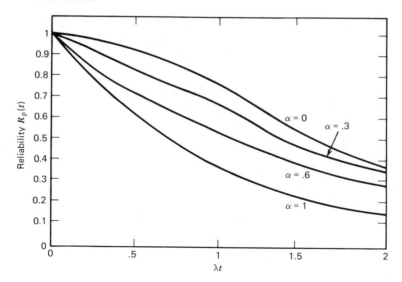

Figure 4.26 A three-parallel-units reliability plot.

cause failure on the parallel system. As the value of α increases, the reliability of the parallel system decreases.

The parameter α takes values from zero to one. At $\alpha=0$, the modified parallel network simply acts as an ordinary parallel network; however, at $\alpha=1$ the modified redundant parallel system just acts as a single unit. What it means is that all the system failures are common-cause failures.

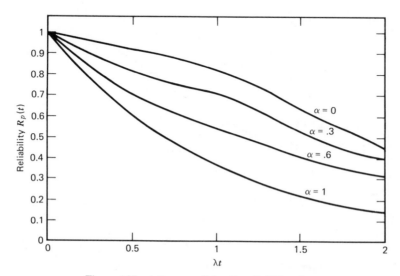

Figure 4.27 A four-parallel-units reliability plot.

The system hazard rate can be obtained from

$$\lambda_p(t) = -\frac{1}{R(t)}\frac{dR(t)}{dt} \tag{4.130}$$

The modified parallel system hazard rate $\lambda_p(t)$ is derived by substituting (4.129) into (4.130):

$$\lambda_p(t) = \alpha\lambda + n\lambda(1-\alpha)\left\{\frac{\gamma-1}{\gamma^n-1}\right\} \tag{4.131}$$

where

$$\gamma = \frac{1}{1-e^{-(1-\alpha)\lambda t}}$$

The hazard rate plot is shown in Figure 4.28. The MTTF can be obtained from

$$\text{MTTF} = \int_0^\infty R(t)\,dt \tag{4.132}$$

The modified parallel system MTTF is obtained by substituting (4.129)

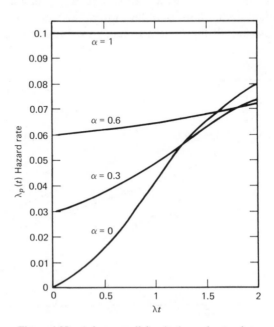

Figure 4.28   A four-parallel-units hazard rate plot.

into (4.132), that is,

$$\text{MTTF} = \sum_{j=1}^{n} \frac{(-1)^{j+1}\binom{n}{j}}{\lambda\{j-(j-1)\beta\}} \qquad (4.133)$$

Example 11. Using the following known data for α, λ, and t. Compute the reliability of a two identical units parallel system:

$$\lambda = 0.001 \text{ failure/hour}$$

$$\alpha = 0.071$$

$$t = 200 \text{ hours}$$

SOLUTION. The reliability of the two identical units parallel system subject to common-cause failures $= 0.95769$. The reliability of the two units parallel system subject to independent failures only $= 0.96714$.

k-*out-of*-n *System.* The modified identical units k-out-of-n system has a hypothetical unit for the common-cause failure connected in series with the independent failure mode k-out-of-n units. The series-connected hypothetical unit represents the system or unit common-cause failures. A failure associated with this hypothetical unit will cause the overall system failure. The modified k-out-of-n identical units system reliability, R_{kn}, can be obtained from

$$R_{kn} = \left\{ \sum_{r=k}^{n} \binom{n}{r} R_1^r (1-R_1)^{n-r} \right\} R_2 \qquad (4.134)$$

where $R_1 = $ the unit independent failure mode reliability
$R_2 = $ the k-out-of-n identical units system common-cause failure reliability

For the constant failure rates λ_1 and λ_2 from (4.126) and (4.127), (4.134) can be rewritten as

$$R_{kn}(t) = \sum_{r=k}^{n} \binom{n}{r} e^{-r(1-\alpha)\lambda t} \{1 - e^{-(1-\alpha)\lambda t}\}^{n-r} e^{-\alpha\lambda t} \qquad (4.135)$$

The graphical plots of (4.135) for 2-out-of-3 units, 2-out-of-4 units, and 3-out-of-4 units are shown in Figures 4.29, 4.30, and 4.31, respectively. As the value of α increases, the system reliability decreases for a small value of λt, as can be verified from Figures 4.29, 4.30, and 4.31.

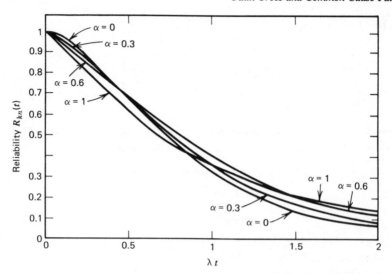

Figure 4.29 A 2-out-of-*n* units reliability plot.

The *k*-out-of-*n* system hazard rate $\lambda_{kn}(t)$ and MTTF can be obtained by substituting (4.135) into (4.130) and (4.132), respectively, that is,

$$\lambda_{kn}(t) =$$

$$-\frac{\left\{ \sum_{r=k}^{n} \binom{n}{r} \left[r\alpha - r - \alpha)\lambda \right] \theta\left[\eta^{(n-r)} \right] + \theta\lambda(n-r)(1-\alpha)\eta^{(n-r-1)}(1-\eta) \right\}}{\sum_{r=k}^{n} \binom{n}{r} \theta \eta^{(n-r)}}$$

$$(4.136)$$

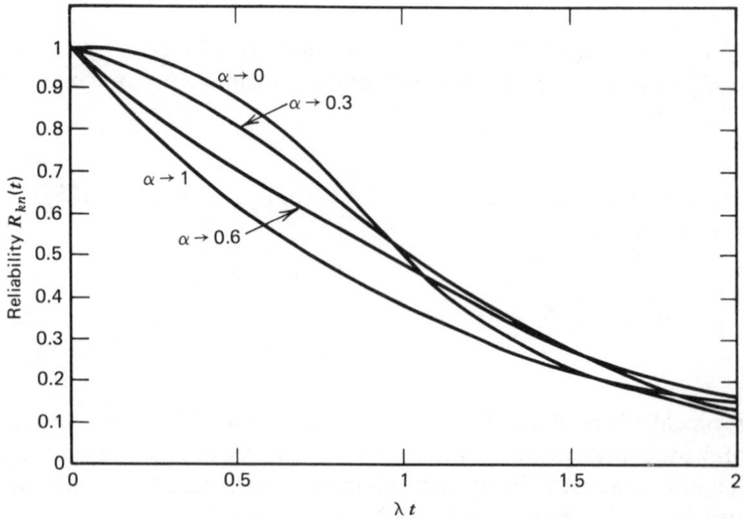

Figure 4.30 A 2-out-of-4 units reliability plot.

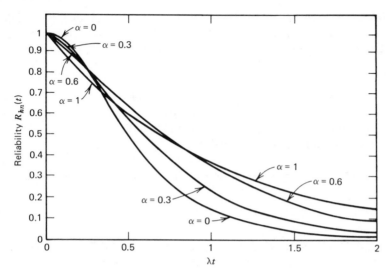

Figure 4.31 A 3-out-of-4 units reliability plot.

where

$$\eta = \left\{1 - e^{-(1-\beta)\lambda t}\right\}$$

$$\theta = e^{(r\alpha - r - \alpha)\lambda t}$$

and

$$\text{MTTF} = \sum_{r=k}^{n} \binom{n}{r} \left[\frac{1}{r - r\alpha + \alpha} - \frac{(n-r)}{(r - r\alpha + 1)\lambda} + \frac{(n-r)(n-r-1)}{2!(r - r\alpha - \alpha + 2)\lambda} \right.$$

$$\left. - \frac{(n-r)(n-r-1)(n-r-2)}{3!} \frac{1}{(r + 3 - r\alpha - 2\alpha)\lambda} + \cdots \right]$$

$$(4.137)$$

Example 12. For the following given hypothetical values of λ, t, and α calculate the system reliability of a 2-out-of-3 units system:

$$\lambda = 0.0005 \text{ failure/hour}$$

$$\alpha = 0.3$$

$$t = 200 \text{ hours}$$

From (4.135) the reliability of a system with common-cause failures was in the order of 0.95772 as compared with the system reliability, 0.97455, with no common cause failures.

Parallel-Series Network.   This system is composed of independent failure mode, identical units, and paths with a hypothetical common-cause failure unit. The modified parallel-series network reliability, R_{ps}, can be obtained from

$$R_{ps} = \{1 - (1 - R_1^m)^n\} R_2 \tag{4.138}$$

where m = the number of identical units in a path
n = the number of identical paths

For the constant failure rates λ_1 and λ_2 the above equation becomes

$$R_{ps}(t) = \left[1 - (1 - e^{-n(1-\alpha)\lambda t})^m\right] e^{-\alpha\lambda t} \tag{4.139}$$

The parallel-series network hazard rate $\lambda_{ps}(t)$ and MTTF can be obtained by substituting (4.139) into (4.130) and (4.132) as follows:

$$\lambda_{ps}(t) = \alpha\lambda + mn(1-\alpha)\lambda \frac{(\gamma-1)}{(\gamma^m-1)} \tag{4.140}$$

where $\lambda = 1/[1 - e^{-n(1-\gamma)\lambda t}]$ and

$$\text{MTTF} = \frac{\sum\limits_{j=1}^{n} \binom{m}{j}(-1)^{j+1}}{\{\lambda\alpha + n\lambda(j-\alpha j)\}} \tag{4.141}$$

A Bridge Network.   This system is composed of an independent failure mode identical units bridge network in series with a hypothetical common-cause failure unit for the bridge structure. If the hypothetical common-cause failure unit fails, the overall system fails. The modified bridge network reliability [127] can be obtained from

$$R_b = \{1 - 2(1 - R_1)^5 + 5(1 - R_1)^4 - 2(1 - R_1)^3 - 2(1 - R_1)^2\} R_2 \tag{4.142}$$

where R_b is the reliability of the bridge network subject to common-cause failures.

For the constant failure rates λ_1 and λ_2 from (4.126) and (4.127), (4.142) can be rewritten as

$$R_b(t) = \left[1 - 2(1 - e^{-At})^5 + 5(1 - e^{-At})^4 - 2(1 - e^{-At})^3\right.$$

$$\left. - 2(1 - e^{At})^2\right] e^{-\beta\lambda t} \tag{4.143}$$

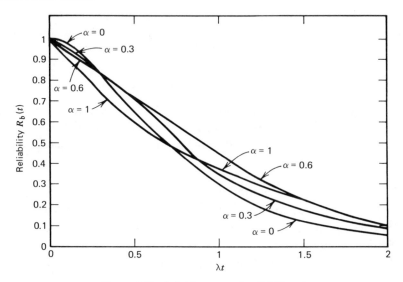

Figure 4.32 A bridge network reliability plot.

where $A=(1-\alpha)\lambda$. The reliability plots of (4.143) are shown in Figure 4.32 for the varying values of parameter α. For the small value of λt, the bridge network reliability decreases as the value of parameter α increases.

The bridge network hazard rate, $\lambda_b(t)$ and the MTTF can be obtained by substituting (4.143) into (4.130) and (4.132), respectively:

$$\lambda_b(t)=\beta\lambda+A(-8\pi^5+25\pi^4-24\pi^3+4\pi^2+4\pi)$$

$$+\frac{-2\pi^5+5\pi^4-2\pi^3-2\pi^2}{1-2\pi^5+5\pi^4-2\pi^3-2\pi^2} \qquad (4.144)$$

where $\pi=(1-e^{-At})$ and

$$\text{MTTF}=\frac{2}{(2-\alpha)\lambda}+\frac{2}{(3-2\alpha)\lambda}+\frac{5}{(4-3\alpha)\lambda}+\frac{2}{(5-4\alpha)\lambda} \qquad (4.145)$$

Example 13. Suppose an identical units bridge network has the following known values for its parameter λ and α. Calculate the bridge reliability for 200 hours, that is,

$$\lambda=0.0005 \text{ failure/hour}$$

$$\alpha=0.3$$

$$t=200 \text{ hours}$$

SOLUTION. The reliability of the bridge network subject to common-cause failures was in order of 0.96 [from (4.143)] as compared to the independent failure mode bridge reliability of 0.984 (i.e., for $\alpha = 0$).

4.14.2 A Common Cause Failure Availability Model

To perform meaningful reliability analysis of a two nonidentical units system with common-cause failures, we present a model taken from reference 92. The following were assumed to develop this mathematical model:

1. Common-cause and other failures are statistically independent.
2. Common-cause failures can only occur with more than one unit.
3. If either one of the active redundant units fails, the unit is repaired. In addition, when both units fail, the system is repaired.
4. The common-cause unit failure and repair rates are constant.

When both units are failed, repair is dependent on the following three cases:

Case 1. The failed component replacements, repair facilities, and skilled craftsmen are available to repair both units.

Case 2. The failed component replacements, repair facilities, and skilled craftsmen are available to repair one unit only.

Case 3. Neither (2) or (3) is applicable due to nonavailability of the failed components replacements, tools, or skilled craftsmen. Furthermore, it may be queuing at a repair facility.

In Case 1 both units can be repaired simultaneously; however, in Case 2 only one unit can be repaired at a time. For the last and final case (3) the units can only be repaired at the availability of the craftsmen replacements for failed components.

The following notations and abbreviations were used to formulate this availability model:

$P_0(t)$ = probability at time t, both units are operational
$P_1(t)$ = probability at time t, the unit 1 has failed and unit 2 is operational
$P_2(t)$ = probability at time t, the unit 2 has failed and unit 1 is operational
$P_3(t)$ = probability at time t, the units 1 and 2 have failed
$P_4(t)$ = probability at time t, the failed component replacements and repairmen are available to repair both units
λ_i = constant failure rate of units 1 and 2, respectively, for $i = 1, 2$
μ_i = constant repair rate of units 1 and 2, respectively, for $i = 1, 2$
μ_3 = constant repair rate of units 1 and 2

α = constant rate of repairmen availability and components replacements

β = constant common-cause failure rate

t = time

Mathematical Model. The system of first-order differential equations [129] associated with Figure 4.33 are

$$P_0'(t) = -\left(\sum_{i=1}^{2} \lambda_i + \beta\right) P_0(t) + \sum_{i=1}^{3} P_i(t)\mu_i + P_4(t)\mu_3 \qquad (4.146)$$

$$P_1'(t) = -(\lambda_2 + \mu_1)P_1(t) + P_3(t)\mu_2 + P_0(t)\lambda_1 \qquad (4.147)$$

$$P_2'(t) = -(\lambda_1 + \mu_2)P_2(t) + P_0(t)\lambda_2 + P_3(t)\mu_1 \qquad (4.148)$$

$$P_3'(t) = -\left(\sum_{i=1}^{3} \mu_i + \alpha\right) P_3(t) + \sum_{i=1}^{2} P_i(t)\lambda_{(3-i)} + P_0(t)\beta \qquad (4.149)$$

$$P_4'(t) = -\mu_3 P_4(t) + P_3(t)\alpha \qquad (4.150)$$

At $P_0(0) = 1$ other initial condition probabilities are equal to zero, where the prime represents differentiation with respect to time t.

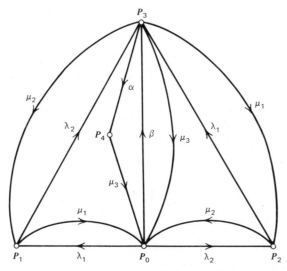

Figure 4.33 A common-cause failure availability model.

The following steady-state equations obtained from (4.146)–(4.150) by setting the derivatives with respect to time t equal to zero:

$$-\left(\sum_{i=1}^{2}\lambda_i+\beta\right)P_0+\sum_{i=1}^{3}P_i\mu_i+P_4\mu_3=0 \qquad (4.151)$$

$$-(\lambda_2+\mu_1)P_1+P_3\mu_2+P_0\lambda_1=0 \qquad (4.152)$$

$$-(\lambda_1+\mu_2)P_2+P_0\lambda_2+P_3\mu_1=0 \qquad (4.153)$$

$$-\left(\sum_{i=1}^{3}\mu_i+\alpha\right)P_3+\sum_{i=1}^{2}P_i\lambda_{(3-i)}+P_0\beta=0 \qquad (4.154)$$

$$-\mu_3P_4+P_3\alpha=0 \qquad (4.155)$$

$$\sum_{i=0}^{4}P_i-1=0 \qquad (4.156)$$

Solving the above system of simultaneous equations yields

$$P_0=\left[\theta\left\{1+\frac{\mu_1(\lambda_2+\mu_1)}{\mu_2(\lambda_1+\lambda_1+\mu_2)}+\frac{\lambda_2+\mu_1}{\mu_2}+\frac{\alpha(\lambda_2+\mu_1)}{\mu_2\mu_3}\right\}\right.$$

$$\left.+\frac{\lambda_2}{\lambda_1+\mu_2}-\frac{\mu_1\lambda_1}{(\lambda_1+\mu_2)\mu_2}-\frac{\lambda_1}{\mu_2}-\frac{\alpha\lambda_1}{\mu_2\mu_3}+1\right]^{-1} \qquad (4.157)$$

where

$$\theta=\frac{P_1}{P_0}$$

$$(P_1/P_0)=\left[\frac{\lambda_1\mu_3+\alpha\lambda_1}{\mu_2}-\left(\frac{\mu_2\lambda_2}{\lambda_1+\mu_2}\right)+\left(\frac{\mu_1\lambda_1}{\lambda_1+\mu_2}\right)+\lambda_1+\lambda_2+\beta\right]$$

$$\times\left[\frac{\mu_3(\lambda_2+\mu_1)+\alpha(\lambda_2+\mu_1)}{\mu_2}+\left(\frac{\mu_1(\lambda_2+\mu_1)}{\lambda_1+\mu_2}\right)+\mu_1\right]^{-1} \qquad (4.158)$$

$$P_1=\theta P_0 \qquad (4.159)$$

$$P_2=\left[\frac{\lambda_2}{\lambda_1+\mu_2}-\frac{\mu_1\lambda_1}{\mu_2(\lambda_1+\mu_2)}\right]P_0+\frac{\mu_1(\lambda_2+\mu_1)P_1}{\mu_2(\lambda_1+\mu_2)} \qquad (4.160)$$

$$P_3=\frac{(\lambda_2+\mu_1)P_1}{\mu_2}-\frac{\lambda_1P_0}{\mu_2} \qquad (4.161)$$

$$P_4=\frac{\alpha P_1(\lambda_2+\mu_1)}{\mu_2\mu_3}-\frac{\alpha\lambda_1}{\mu_2\mu_3}P_0 \qquad (4.162)$$

The steady-state system availability can be obtained from

$$\text{system availability} = \sum_{i=0}^{2} P_i \qquad (4.163)$$

4.14.3 A 1-Out-Of-N: G System With Duplex Elements

This model incorporates stand-by duplex unit replacements and common-cause failures [91]. When the operational duplex system (contains two statistically identical units) fails, it is replaced by one of the $(N-1)$ standby duplex systems. Furthermore, this model incorporates a possibility (i.e., to replace the failed system) that the repairmen or special repair tools may be available or, alternatively, not available at the time of the operational system failure. This type of situation occurs at a nuclear plant where a duplex system is replaced only when both units fail.

The following were assumed to develop this model:

1. A duplex system has two statistically identical units. All but one of the duplex systems are cold standbys (units cannot fail).
2. Common-cause and other failures are statistically independent.
3. Operational system is replaced only when both units fail.
4. Operational units are independently identically distributed (i.i.d.), except for common-cause failures.
5. A failed system is restored as good as new.
6. Cold standby systems; standby units cannot fail.
7. Failed duplex systems are never repaired.
8. When a system fails, two different possibilities are considered to replace it with one of the standbys: (a) repairmen and special repair tools are available; (b) repairmen and special repair tools are not available.

The notation for Figure 4.34 is as follows:

n = total number of system states
i = state of the system: both units are good, $i = 0, 4, 8, \ldots, (n-2)$; one unit is good, one unit is bad, $i = 1, 5, 9, \ldots, (n-1)$; both units are bad, no waiting, $i = 2, 6, 10, \ldots, n$; repairmen or special repair tool waiting state, $i = 3, 7, 11, \ldots, (n-3)$
λ = constant failure (hazard) rate of a single unit
β = constant common-cause failure (hazard) rate of the duplex system
α = constant repairmen or special repair tool availability (hazard) rate
μ_j = constant and replacement (hazard) rate of the failed duplex system when repairmen and special repair tools are available (for $j=1$); or not available (for $j=2$)

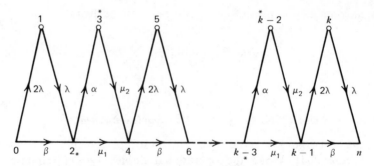

Figure 4.34 Transition diagram of system. The star denotes down states.

The equations for the Figure 4.34 model [129] are:

$$P_0'(t) = -(2\lambda + \beta)P_0(t) \tag{4.164}$$

$$P_k'(t) = P_{k-1}(t)2\lambda - P_k(t)\lambda \tag{4.165}$$

$$P_{k-3}'(t) = P_{k-5}(t)\beta + \lambda P_{k-4}(t) - (\mu_1 + \alpha)P_{k-3}(t) \tag{4.166}$$

$$P_{k-2}'(t) = P_{k-3}(t)\alpha - \mu_2 P_{k-2}(t) \tag{4.167}$$

$$P_{k-1}'(t) = P_{k-3}(t)\mu_1 + \mu_2 P_{k-2}(t) - (2\lambda + \beta)P_{k-1}(t) \tag{4.168}$$

$$\vdots$$

$$P_n'(t) = P_k(t)\lambda + P_{k-1}(t)\beta \tag{4.169}$$

The above equations are valid for $k = 5, 9, 13, \ldots, (n-1)$.

$$P_i(t) = 1 \qquad \text{for} \quad i = 0$$

$$= 0 \qquad \text{for all other } i$$

The prime denotes differentiation with respect to time t.

$$n \equiv (4N - 2) \qquad \text{for} \quad N \geqslant 2 \tag{4.170}$$

The Laplace transforms of the end result are

$$P_0(s) = \frac{1}{s + 2\lambda + \beta} \tag{4.171}$$

$$P_k(s) = \frac{P_{k-1}(s)2\lambda}{s + \lambda} \tag{4.172}$$

$$P_{k-2}(s) = \frac{P_{k-3}(s)\alpha}{s + \mu_2} \tag{4.173}$$

$$P_{k-3}(s) = \frac{P_{k-5}(s)\beta + P_{k-4}(s)\lambda}{s + \mu_1 + \alpha} \qquad (4.174)$$

$$P_{k-1}(s) = \frac{P_{k-3}(s)\mu_1 + P_{k-2}(s)\mu_2}{s + 2\lambda + \beta} \qquad (4.175)$$

$$\vdots$$

$$P_n(s) = \frac{P_k(s)\lambda + P_{k-1}(s)\beta}{s} \qquad (4.176)$$

To obtain the time domain solution, one should transform (4.171)–(4.176) for the known value of N.

4.14.4 A 4-Unit Redundant System with Common-Cause Failures

This mathematical model represents a 4-identical-unit system with common-cause failures [87] where system repair times are arbitrarily distributed. Therefore, the supplementary variable technique [123, 125, 126] is used to develop equations for the model.

The following were assumed to develop this mathematical model:

1. Common-cause and other failures are statistically-independent.
2. Common-cause failures can only occur with more than one unit.
3. Units are repaired only when the system fails. A failed system is restored to like-new.
4. System repair times are arbitrarily distributed.

The transition diagram is shown in Figure 4.35.

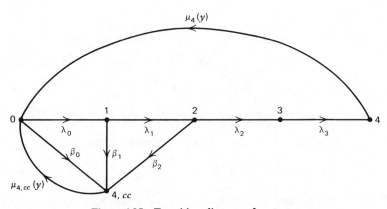

Figure 4.35 Transition diagram of system.

The following notation was used to develop model equations

i = state of the unfailed system: number of failed units, $i=0,1,2,3$

j = state of the failed system, $j=4$ means failure not due to a common cause; $j=4$, cc means failure due to a common-cause

$P_i(t)$ = probability that system is in unfailed state i, at time t

$p_j(y,t)$ = probability density (with respect to repair time) that the failed system is in state j and has an elapsed repair time of y

$\mu_j(y)$, $q_j(y)$ = repair rate (a hazard rate) and pdf of repair time when system is in state j and has an elapsed repair time of y

β_i = constant common-cause failure rate of the system when in state i; $\beta_3=0$

λ_i = constant failure rate of a unit, for other than common-cause failures, when the system is in state i; $i=0,1,2,3$

s = Laplace transform variable

The equations for the model are

$$\frac{dP_0(t)}{dt} + (\lambda_0+\beta_0)P_0(t)$$

$$= \int_0^\infty p_4(y,t)\mu_4(y)\,dy + \int_0^\infty p_{4,cc}(y,t)\mu_{4,cc}(y)\,dy$$

$$(4.177)$$

$$\frac{dP_i(t)}{dt} + (\lambda_i+\beta_i)P_i(t) - \lambda_{i-1}P_{i-1}(t) = 0$$

$$i=1,2,3 \qquad \beta_3=0 \quad (4.178)$$

$$\frac{\partial p_j(y,t)}{\partial t} + \frac{\partial p_j(y,t)}{\partial y} + \mu_j(y)p_j(y,t) = 0$$

$$j=4 \quad \text{or} \quad 4,cc \quad (4.179)$$

$$p_4(0,t) = \lambda_3 P_3(t) \qquad\qquad (4.180)$$

$$p_{4,cc}(0,t) = P_0(t)\beta_0 + P_1(t)\beta_1 + P_2(t)\beta_2 \qquad (4.181)$$

$$P_i(0)=1 \quad \text{for} \quad i=0 \quad \text{otherwise} \quad P_i(0)=0$$

$$p_j(y,0)=0 \qquad \text{for all} \quad j \qquad\qquad (4.182)$$

Solving and setting up, similar to the above equations are presented in reference 125.

The Laplace transforms of the solution are

$$P_0(s) = \left[s + \lambda_0 + \beta_0 - \left(\beta_0 + \frac{\beta_1}{A_1} + \frac{\beta_2}{A_2} \right) G_{4,cc}(s) - \frac{G_4(s)}{A_3} \right]^{-1} \quad (4.183)$$

$$G_j(s) \equiv \int_0^\infty \exp(-sy) q_j(y)\, dy \qquad \text{for} \quad j=4 \quad \text{or} \quad 4,cc$$

$$A_1 \equiv \frac{s + \lambda_1 + \beta_1}{\lambda_0}$$

$$A_2 \equiv \frac{A_1(s + \lambda_2 + \beta_2)}{\lambda_1}$$

$$A_3 \equiv \frac{A_2(s + \lambda_3)}{\lambda_2}$$

$$P_i(s) = \frac{P_0(s)}{A_i} \qquad \text{for} \quad i = 1, 2, 3 \tag{4.184}$$

$$P_4(s) = \frac{\lambda_3 P_3(s)[1 - G_4(s)]}{s} \tag{4.185}$$

$$P_{4,cc}(s) = \left[\sum_{i=0}^{2} \beta_i P_i(s) \right] \frac{1 - G_{4,cc}(s)}{s} \tag{4.186}$$

To obtain time domain solution of the above equations, one should substitute the Laplace transform of the repair times density functions for $G_4(s)$ and $G_{4,cc}(s)$ and then take inverse Laplace transforms of (4.183)–(4.186).

REFERENCES

This section presents a bibliography on fault trees and common-cause failures as well as some miscellaneous references related to the subject of interest. It is expected that this selected bibliography, in this important and fast developing area, will be of considerable interest to the readers.

Fault Trees

1. Barlow, R. E. and F. Proschan, *Statistical Theory of Reliability and Life Testing—Probability Models*, Holt, Rinehart and Winston, New York, 1975.
2. Barlow, R. E., J. B. Fussell, and N. D. Singpurwalla, *Reliability and Fault Tree Analysis*. SIAM, Philadelphia, 1975.

3. Green, A. E. and A. J. Bourne, *Reliability Technology*. Wiley-Interscience, London, 1972.

4. Aggarwal, K. K., "Comments on 'On the Analysis of Fault Trees,'" *IEEE Trans. Reliab.*, **R-25**, 126–127 (1976).

5. Apostolakis, G. and Y. T. Lee, "Methods for the Estimation of Confidence Bounds for the Top-Event Unavailability of Fault Trees," *Nucl. Eng. Design*, **41**, 411–419 (1977).

6. Barlow, R. E. and H. E. Lambert, "Introduction to Fault Trees Analysis." In: *Reliability and Fault Tree Analysis*, SIAM, Philadelphia, 1975, pp. 7–37.

7. Bass, L., H. W. Wynholds, and W. R. Poterfield, "Fault Tree Graphics," *Proceedings of the Annual Reliability Maintainability Symposium*, IEEE, New York, 1975, pp. 292–297.

8. Bazovsky, I., "Fault Trees, Block Diagrams and Markov Graphs," *Proceedings of the Annual Reliability Maintainability Symposium*, IEEE, New York, 1977, pp. 134–141.

9. Bengiamin, N. N., B. A. Brown, and K. F. Schenk, "An Efficient Algorithm for Reducing the Complexity of Computation in Fault Tree Analysis," *IEEE Trans. Nucl. Sci.* NS-23, pp. 1442–1446 (1976).

10. Bennetts, R. G., "On the Analysis of Fault Trees," *IEEE Trans. Reliab.* **R-24** (3), 175–185 (1975).

11. Browning, R. L., "Analyzing Industrial Risks," *Chem. Eng.*, **76**, 109–114 (Oct. 1969).

12. Burdick, G. R., "COMCAN—A Computer Code for Common-Cause Analysis," *IEEE Trans. Reliab.*, **R-26**, 100–102 (1977).

13. Carnino, A., "Safety Analysis Using Fault Trees," *NATO Advanced Study Institute on Generic Techniques of System Reliability Assessment*, Nordhoff, Leiden, Netherlands, 1974.

14. Chatterjee, P., "A Method to Reduce the Cost of Analysis," In: *Reliability and Fault Tree Analysis*, SIAM, Philadelphia, 1975, pp. 101–129.

15. Crosetti, P. A., "Fault Tree Analysis with Probability Evaluation," *IEEE Trans. Nucl. Sco.*, **18**, 465–471 (1971).

16. Crosetti, P. A. and R. A. Bruce, "Commercial Application of Fault Tree Analysis," *Proceedings of the Annual Reliability Maintainability Symposium*, IEEE, New York, 1970.

17. Cummings, G. E., "Application of the Fault Tree Technique to a Nuclear Reactor Containment System," In: *Reliability and Fault Tree Analysis*, SIAM, Philadelphia, 1975.

18. Danzeisen, R. N., J. A. Mateyka, and D. W. Weiss, "A Reliability and Safety Analysis of Automotive Vehicles," *Proceedings of the Symposium on Reliability*, 1971, p. 150–151. IEEE, New York.

19. Dhillon, B. S. " A Modification of Fault Tree AND Gate," *Microelectron. Reliab.*, **15**, 625–626 (1976).

20. Dhillon, B. S. and C. Singh, "On Fault Trees and Other Reliability Evaluation Methods," *Microelectron. Reliab.*, **19** (1/2), 57–64 (1979).

21. Dhillon, B. S. and C. Singh, "Bibliography of Literature on Fault Trees," *Microelectron. Reliab.*, **17**, 501–503 (1978).

22. Dhillon, B. S., C. L. Proctor, and A. Kothari, On repairable component fault tree, *Proceedings of the Annual Reliability and Maintainability Symposium*, (1979). IEEE, New York.

23. Eagle, K. H., "Fault Tree and Reliability Analysis Comparison," *Proceedings of the Symposium on Reliability*, (1969). IEEE, New York.

24. Evans, R. A., "Fault-Trees and Cause-Consequence Charts," *IEEE Trans. Reliab.*, **23**, 1 (1974).

25. Fleming, K. N., G. W. Hannaman, "Common Cause Failure Considerations in Predicting HTGR Cooling System Reliability," *IEEE Trans. Reliab.*, **R-25**, 171–177 (1976).

26. Fussell, J. B., "Fault Tree Analysis—Concepts and Techniques," *Proceedings of the NATO Advanced Study Institute on Generic Techniques of System Reliability Assessment*, Nordhoff, Leiden, Netherlands, 1975

27. Fussell, J. B., "Fault Tree Analysis—Concepts and Techniques," *Proceedings of the NATO Advanced Study Institute on Generic Techniques of System Reliability Assessment*, Nordhoff, Leiden, Netherlands, 1975.

28. Fussell, J. B., "Computer Aided Fault Tree Construction for Electrical System." In: *Reliability and Fault Tree Analysis*, SIAM, Philadelphia, 1975.

29. Fussell, J. B., "A Formal Methodology for Fault Tree Construction," *Nucl. Sci. Engng*, **52**, 421–432 (1973).

30. Fussell, J. B., "How to Hand-Calculate System Reliability Characteristics," *IEEE Trans. Reliab.*, **R-24**, 169–174 (1975).

31. Fussell, J. B., E. F. Aber, and R. G. Rahl, "On the Quantitative Analysis of Property—AND Failure Logic," *IEEE Trans. Reliab.*, **R-26**, 324–326 (1977).

32. Fussell, J. B., G. J. Powers, and R. G. Bennetts, "Fault Trees—A State of the Art Discussion," *IEEE Trans. Reliab.*, **R-23**, 51–55 (1974).

33. Fussell, J. B. and W. E. Vesely, "A New Methodology for Obtaining Cut Sets for Fault-Trees," *Trans. Am. Nucl. Soc.*, **15**, 262–263 (1972).

34. Garribba, S., "Efficient Construction of Minimal Cut Set from Fault Trees," *IEEE Trans. Reliab.*, **R-26**, 88–94 (1977).

35. Garrick, B. J., "Principles of Unified Systems Safety Analysis," *Nucl. Eng. Design*, **13**, 245–321 (1970).

36. Gopal, K. and J. S. Gupta, "On the Analysis of Fault Trees—Some Comments," *IEEE Trans. Reliab.*, **R-26**, 14–15 (April 1977).

37. Haasl, D. F., "Advanced Concepts in Fault Tree Analysis," *System Safety Symposium*, (1965). (Available from the University of Washington Library, Seattle.)

38. Hannum, W. H., F. X. Gavigan, D. E. Emon, "Reliability and Safety Analysis Methodology in the Nuclear Programs of ERDA," *IEEE Trans. Reliab.*, **R-25**, 140–146 (1976).

39. Henley, E. J., "Systems Analysis by Sequential Fault Trees," *Microelectronics and Reliab.*, **15**, 247–248 (1976).

40. Henser, F. W., "Reliability Analysis of Reactor Systems," *Proceedings of the Annual Symposium on Reliability*, 1970, pp. 135–145. IEEE, New York.

41. Hiltz, P. A., "The Fundamentals of Fault-Tree Analysis," Government-Industry Data Exchange Program (GIDEP), Report No. 347.40.00.00-F1-38 (C2300). Available from the GIDEP Operations Center, Corona, CA 91720, 1965.

42. Human, C. L., "The Graphical FMECA", *Proceedings of the Annual Reliability and Maintainability Symposium*, IEEE, New York, 1975, pp. 298–303.

43. Lambert, H. E., "Measures of Importance of Events and Cut Sets in Fault Trees," In: *Reliability and Fault Tree Analysis*, SIAM, Philadelphia, 1975, pp. 77–101.

44. Lambert, H. E. and G. Yadigaroglu, "Fault Trees for Diagnosis of System Fault Conditions," *Nucl. Sci. Eng.*, **62**, 20–34 (1977).

45. Lapp, S. A. and G. J. Powers, "Computer-Aided Synthesis of Fault Trees," *IEEE Trans. Reliab.*, **R-26**, 2–13 (1977).

46. Levine, S. and W. E. Vesely, "Important Event-Tree and Fault-Tree Considerations in the Reactor Safety Study," *IEEE Trans. Reliab.*, **R-25**, 132–139 (1976).

47. Michels, J. M., "Computer Evaluation of the Safety Fault Tree Model," System Safety Symposium, (1965). (Available from the University of Washington Library, Seattle.)

48. Murchland, J. D., "Comments on 'A Time Dependent Methodology for Fault Tree Evaluation,'" *Nucl. Eng. Design*, **22**, 167–172 (1972).

49. Nagel, P. M., "A Monte Carlo Method to Compute Fault Tree Probabilities," *System Safety Symposium*, (1965). (Available from University of Washington Library, Seattle.) 1965.

50. Neilsen, D. S., O. Platz, and B. Runge, "A Cause Consequence Chart of a Redundant Protection System," *IEEE Trans. Reliab.*, **R-24**, (April 1975). pp 8–13.

51. Neogy, R., "Fault Trees in Ocean Systems," *Proceedings of the American Reliability and Maintainability Symposium*, IEEE, New York, 1975, pp. 280–285.

52. Neuman, C. P. and N. M. Bonhomme, "Evaluation of Maintenance Policies using Markov Chains and Fault Tree Analysis," *IEEE Trans. Reliab.*, **24**, (1975). pp 37–44.

53. Nieuwhof, G. W. E., "An Introduction to Fault Tree Analysis with Emphasis on Failure Rate Evaluation," *Microelectron. Reliab.*, **14**, 105–119 (1975).

54. Nieuwhof, G. W. E., "Unavailability Logic Tree Analysis," *Third Annual Reliability and Energy Conference on Power Apparatus, Montreal*, IEEE, New York, 1976, pp. 79–83.

55. Phibbs, E. and S. H. Kuwamoto, "An Efficient Map Method for Processing Multistate Logic Trees," *IEEE Trans. Reliab.*, **R-21**, (1972). pp. 93–98.

56. Powers, G. J. and F. C. Tompkins, "Fault Tree Synthesis for Chemical Processes," *Amer. Inst. Chem. Eng. J.*, **20**, 376–387 (1974).

57. Powers, G. J., F. C. Tompkins, and S. A. Lapp, "A Safety Simulation Language for Chemical Processes: A Procedure for Fault Tree Synthesis," In: *Reliability and Fault Tree Analysis*, SIAM, Philadelphia, 1975.

58. Roush, S. L. and F. J. Schilagi, "Fault Tree Approach to Organization Design," *Proceedings of the Annual Reliability Maintainability Symposium*, IEEE, New York, 1971.

59. Rubel, P., "BONSAI: Cultivating th Logic Tree for Reactor Safety," *Proceedings of the Annual Reliability and Maintainability Symposium*, (Available from the IEEE). 1975.

60. Rubel, P., "Tiger in the Fault Tree Jungle," *Proceedings of the Seventh Annual Modelling Simulation Conference*, Pittsburgh, Instrument Society of America, 1975, pp. 1071–1086. 400 Stanwix St., Pittsburgh, Pennsylvania 15222, USA.

61. Reactor Safety Study, WASH-1400 (NUREG-75) National Technical Information Service, Springfield, VA, 22161, (October 1976).

62. Salvatori, R., "Systematic Approach to Safety Design and Evaluation," *IEEE Trans. Nucl. Sci.*, **18**, (February 1971).

63. Sarver, S. J., "Reliability Evaluation of a Containment for Cooler System," *Proceedings of the Annual Reliability and Maintainability Symposium*, IEEE, New York, 1975, pp. 154–162.

64. Schneeweiss, W. G., "Calculating the Probability of Boolean Expression Being 1," *IEEE Trans. Reliab.*, **26**, (1977).

65. Schroder, R. J., "Fault Tree for Reliability Analysis," *Proceedings of the Annual Symposium Reliability*, IEEE, New York, 1970.

66. Semanderes, S. N., "Elraft, a Computer Program for Efficient Logic Reduction Analysis of Fault Trees," *IEEE Trans. Nucl. Sci.*, **18**, (February 1971).

67. Shooman, M. L., "The Equivalence of Reliability Diagrams and Fault-Tree Analysis," *IEEE Trans. Reliab.*, **R-19**, 74–75 (1970).

68. Vesely, W. E., "A Time-Dependent Methodology for Fault Tree Evaluation," *Nucl. Eng. Design*, **13**, 337–360. (April, 1970).

69. Vesely, W. F., "Reliability and Fault-Tree Applications at the NRTS," *IEEE Trans. Nucl. Sci.*, **18**, 472–480 (February 1971).

70. Virolainen, R., "Unreliability of a Complex System with Parallel Redundancy and Repair," *Nucl. Eng. Design*, **40**, 431–441 (1977).

71. Wheeler, D. B. et al., "Fault Tree Analysis Using Bit Manipulation," *IEEE Trans. Reliab.* **R-26**, 95–99 (June 1977).

72. Worrell, R. B., "Using the Set Equation Transformation System in Fault Tree Analysis," In: *Reliability and Fault Tree Analysis*, SIAM, Philadelphia, 1975.

73. Wynholds, H. W. and R. Poterfield, "Fault-Tree Graphics," In: *Reliability and Fault Tree Analysis*, SIAM, Philadelphia, 1975.

74. Young, J., "Using the Fault Tree Analysis Technique," In: *Reliability and Fault Tree Analysis*, SIAM, Philadelphia, 1975.

Common-Cause Failures

75. Apostolakis, G. E., "The Effect of a Certain Class of Potential Common-Mode Failures on the Reliability of Redundant Systems," *Nucl. Eng. Design*, **36**, 123–133 (1976).

76. Apostolakis, G. E., "On a Certain Class of Potential Common-Mode Failures," *Trans. Am. Nucl. Soc.*, **22**, 476–477 (1975).

77. Apostolakis, G. E., "On the Reliability of Redundant Systems," *Trans. Am. Nucl. Soc.*, **22**, 477–478 (1975).

78. Barlow, R. T., R. A. Hill, D. C. McIntine, and J. F. O'Brien, "Probabilistic Evaluation of Failure Modes Leading to the Mispositioning of Certain ECCS Motor Operated Valves in Westinghouse NSS," *IEEE Trans. Power Appar. Syst.*, **PAS-97**, 358–361 (March/April 1978).

79. Billington, R., T. K. P. Medicherla, and M. S. Sachdev, "Common-Cause Outages in Multiple Circuit Transmission Lines," *IEEE Trans. Reliability*, **27**, 128–131 (1978).

80. Burdick, G. R., "COMCAN—A Computer Code for Common-Cause Analysis," *IEEE Trans. Reliab.*, **R-26**, 100–102 (1977).

81. Cain, D. G., "Closing the Loop Summary-Equipment Qualification," *Proceedings of the 1977 Environmental Technology Conference*. Institute of the Environmental Sciences, IL, 60058, pp. 434–437. (1977).

82. Cate, C. L., D. P. Wagner and J. B. Fussell, "A Computer Aided Approach to Qualitative and Quantitative Common-Cause Failure Analysis for Complex Systems," *Proceedings of the Eighth Pittsburgh Modelling and Simulation Conference*, Instrument Society of America, Pittsburgh, 1977, pp. 25–29.

83. Chelapati, C. V., and R. P. Kennedy, "Probabilistic Assessment of Aircraft Hazard for Nuclear Plants," *Nucl. Eng. Design*, **19**, 333–364 (1972).

84. Chu, B. B. and D. P. Gaver, "A Stochastic Modelling for Common-Mode Failures of Repairable Redundant Systems," *Conference Proceedings*, Edited by J. B. Fussell and G. R. Burdick, SIAM, Philadelphia, 1977.

85. Dhillon, B. S., "A k-out-of-N Three-state Devices System with Common-cause Failures," *Microelectron. Reliab.*, **18**, 447–448 (1978).

86. Dhillon, B. S., "A Modification to Fault Tree "AND" Gate," *Microelectron. Reliab.*, **15**, 625–626 (1976).

87. Dhillon, B. S., "A 4-Unit Redundant System with Common-Cause Failures," *IEEE Trans. Reliab.*, **R-26**, 373–374 (1977).

88. Dhillon, B. S. and C. L. Proctor, "Common-Mode Failure Analysis of Reliability Networks," *Proceedings of the Annual Reliability and Maintainability Symposium*, IEEE, New York, 1977, pp. 404–408.

89. Dhillon, B. S., "Effects of Weibull Hazard Rate on Common-Cause Analysis of Reliability Networks," *Microelectron. Reliability*, **17**, 59–65 (1978).

90. Dhillon, B. S., "Optimal Maintenance Policy for Systems with Common-Cause Failures," *Proceedings of the Ninth Pittsburgh Modelling and Simulation Conference*, Instrument Society of America, Pittsburgh, 1978.

91. Dhillon, B. S., "A 1-out-of-N: G system with Duplex Elements," *IEEE Trans. Reliab.* (in press).

92. Dhillon, B. S., "A Common-Cause Failure Availability Model," *Microelectron. Reliability*, **17**, 583–584 (1978).

93. Dhillon, B. S., "ON Common-Cause Failures—Bibliography," *Microelectron. Reliability*, **18**, 533–534 (1978).

94. Dhillon, B. S., "A 4-Unit Redundant System with Common-Cause Failures," *IEEE Trans. Reliability*, **R-28**, 267 pp. (June 1979).

95. Dhillon, B. S., A. Sambhi, and M. R. Khan, "Common Cause Failure Analysis of a Three-State Device System," *Microelectron. Reliab.*, **19**, 345–348 (1979).

96. Ditto, S. J., "Failures of Systems Designed for High Reliability," *Nucl. Safety*, **8**, 35–37 (Fall 1966).

97. Epler, E. P., "Common-Mode Failure Considerations in the Design of Systems for Protection and Control," *Nucl. Safety*, **11**, 323–327 (Jan.–Feb. 1969).

98. Epler, E. P., "The ORR Emergency Cooling Failures," *Nucl. Safety*, **11**, 323–327 (July–Aug. 1970).

99. Epler, E. P., "Diversity and Periodic Testing in Defense Against Common-Mode Failures," Edited by J. B. Fussell and G. R. Burdick, *Conference Proceedings*, Society for Industrial and Applied Mathematics, Philadelphia, PA 19103, 1977.

100. Evans, R. A., "Statistical Independence and Common-Mode Failures," *IEEE Trans. Reliab.*, **R-24**, 289 (1975).

101. Fleming, K. N., "A Redundant Model for Common Mode Failures in Redundant Safety Systems," *Proceedings of the Sixth Pittsburgh Annual Modelling and Simulation Conference*, Instrument Society of America, Pittsburgh, 1975, pp. 579–581

102. Fleming, K. N., and G. W. Hannaman, "Common Cause Failure Considerations in Predicting HTGR Cooling System Reliability," *IEEE Trans. Reliab.*, **R-25**, 171–177 (1976).

103. Gachot, B., "A Probabilistic Approach to Design for the ECCS of a PWR," *Proceedings of the Annual Reliability and Maintainability Symposium*, 1977, pp. 332–342.

104. Gangloff, W. C., "Common-Mode Failure Analysis is "in"," *Electron. World*, 30–33 (Oct. 1972).

105. Gangloff, W. C., "Common Mode Failure Analysis," *IEEE Trans. Power Apparatus Systems*, **94**, 27–30 (Feb. 1975).

106. Gangloff, W. C. and T. Franke, "An Engineering Approach to Common-Mode Failure Analysis," *Conference on the Development and Application of Reliability Techniques to Nuclear Plants, Liverpool, England* (April 1974). University of Liverpool.

107. Garrick, B. J., W. C. Gekler, and H. P. Pomrehn, "Some Aspects of Protective Systems in Nuclear Power Plants," *IEEE Trans. Nucl. Sci.*, **NS-12**, 22–30 (Dec. 1975).

108. Hayden, K. C., "Common-Mode Failure Mechanisms in Redundant Systems Important to Reactor Safety," *Nucl. Safety*, **17**, 686–693 (1976).

109. Houghton, W. J., V. Joksimovic, and D. E. Emon, "Methods of Probabilistic Safety Analysis for Gas-Cooled Reactors," *Trans. Amer. Nucl. Soc.*, **21**, 210–217 (1975).

110. Jacobs, I. M., "The Common-Mode Failure Study Discipline," *IEEE Trans. Nucl. Sci.*, **NS-17**, 594–598 (1970).

111. Levine, S. and W. E. Vesely, "Important Event-Tree and Fault-Tree Considerations in the Reactor Safety Study," *IEEE Trans. Reliab.*, **R-25**, 132–139 (1976).

112. Leverenz, F. L., E. T. Rumble, and E. Erdmann, "A Dependent-Event Model for Fault Trees," *Trans. Amer. Nucl. Soc.*, **21**, 212–213 (1975).

113. Rankin, J. P., "Sneak-Circuit Analysis," *Nucl. Safety*, **14**, 461–469, (Sept.–Oct. 1973).

114. *Reactor Safety Study*, WASH-1400 (NUREG-75/014)(Oct. 1975), National Technical Information Service, Springfield, VA 22161.

115. Rubel, P., "Tiger in the Fault Tree Jungle," *Proceedings of the Seventh Annual Pittsburgh Modelling and Simulation Conference*, Pittsburgh, Instrument Society of America, 1976, pp. 1071–1082.

116. Taylor, J. R., "A Study of Failure Causes Based on U.S. Power Reactor Abnormal Occurrence Reports," *Reliab. Nucl. Power Plants*, IAEA-SM-195/16 (1975).

117. Vesely, W. E., "Estimating Common-Cause Failure Probabilities in Reliability and Risk Analysis: Marshal-Olkin Specialization," Edited by J. B. Fussell and G. R. Burdick, Society for Industrial and Applied Mathematics, Philadelphia, PA 19103, 1977.

118. Wagner, D. P., C. L. Cate, and J. B. Fussell, "Common-Cause Failure Analysis Methodology for Complex Systems," Edited by J. B. Fussell and G. R. Burdick, *Conference Proceedings*, Society for Industrial and Applied Mathematics, 33 South 17 St., Philadelphia Pennsylvania 19013 U.S.A.

119. Wall, I. B., "Probabilistic Assessment of Flooding for Nuclear Power Plants," *Nucl. Safety*, **15**, 399–408 (1974).

120. Wall, I. B., "Probabilistic Assessment of Aircraft Risk for Nuclear Power Plants," *Nucl. Safety*, **15**, 276–284 (1974).

121. Wilson, J. R. and R. J. Crump, "Computer-Aided Common-Cause Analysis of an LMFBR System," *Trans. Amer. Nucl. Soc.*, **22**, 474–475 (1975).

122. Worrell, R. B. and G. R. Burdick, "Qualitative Analysis in Reliability and Safety Studies," *IEEE Trans. Reliab.*, **R-25**, 164–169 (1976).

Miscellaneous

123. Cox, D. R., "The Analysis of Non-Markovian Stochastic Process by Supplementary Variables," *Proc. Camb. Phil. Soc.*, **51**, 433–441 (1955).

124. *Flow Research Report*, "Risk Analysis Using the Fault Tree Technique," Flow Research, Inc., 1973.

125. Garg, R. C., "Dependability of a Complex System Having Two Types of Components," *IEEE Trans. Reliab.*, **R-12**, 11–15 (Sept. 1963).

126. Gaver, D. P., "Time to Failure and Availability of Paralleled Systems with Repair," *IEEE Trans. Reliab.*, **R-12**, 30–38 (June 1963).

127. Lipp, J. P., "Topology of Switching Elements vs. Reliability," *Trans. IRE Reliab. Quality Control* (June 1957).

128. Ross, S. M., "On the Calculation of Asymptotic System Reliability Characteristics," *Reliability and Fault Tree Analysis*, SIAM, Philadelphia, 1975, pp. 331–350.

129. Shooman, M. L., *Probabilistic Reliability: An Engineering Approach*, McGraw-Hill, New York, 1968.

130. Wolfe, W. A., "Fault Trees Revisited," *Microelectron. Reliab.*, **17**, (1978).

5

Software Reliability

5.1 INTRODUCTION

Computers are finding an ever-increasing number of applications. The expenditure on computer software is increasing faster than on associated hardware. One of the estimates indicates [1] that the annual expenditure of the U.S. Air Force, for example, on computer hardware is $400 million and the corresponding expenditure for software is estimated at $1500 million per year. This ratio of four to one is predicted to rise to nine to one [5, 23]. With expenditures of this magnitude, it is natural that attention should be directed to the proper development of software for computer applications. One area on which considerable emphasis has been placed in recent years is software reliability. This has come primarily with the advent of large and complex hardware-software systems and the use of computers as the heart of real-time applications to control vital and critical functions. The undetected errors can cause system failures with catastrophic results and at the same time the size and complexity has increased making the process of debugging more difficult. Most of the work in the area of software reliability can be divided into the following three categories:

1. Writing correct programs to begin with.
2. Testing the programs to take out the bugs.
3. Modeling of software in an attempt to predict its reliability and possibly study the impact of related parameters.

These three areas are discussed in this chapter. It should be pointed out that software reliability is in no way as highly developed as the discipline of hardware reliability. Several useful concepts have, however, emerged, and considerable work is still under progress.

5.2 HARDWARE AND SOFTWARE

The discipline of hardware reliability is considerably older than that of software. It is, therefore, natural to make a comparison between the two in an effort to apply the large body of knowledge of reliability engineering to software assurance. Several attempts [8, 13] have been made in this direction. It appears that much can be learned from established reliability engineering in organization and control procedures but that there are significant differences when it comes to failure mechanisms. A hardware component is assumed to have failed if its characteristics change beyond the design values either by drift or catastrophic failures. A piece of software, however, does not fail. If a program does not do what it is supposed to do, it is because an error is present. The error has been there and when the segment of the program containing the error is energized, the error becomes manifest. This encountering of error may or may not cause a system to fail. Whereas the hardware undergoes a change at the instant of failure, the software is really the same as it was before the error was discovered.

The hardware reliability of a system can be improved by using two identical components in a redundant manner. Two identical softwares, however, will be of little use in increasing the reliability since the same error will be exercised in both at the same time.

There is an important difference between hardware and software in regard to the relationship between testing and reliability. If the software could be tested for every conceivable input, then theoretically it should never cause system failure. The hardware, on the other hand, could fail even after having been tested in the most exhaustive manner.

A question that may be asked is, "Can the failure behavior of software be regarded as random?" A program basically maps the elements of input space into corresponding elements of output space [11]. A certain subset of the input space would produce incorrect output. If we knew the output behavior for every conceivable input and could predict the future inputs, then we could predict the failures in a deterministic fashion. The properties of a large piece of software, however, may never be known completely, since it is almost impossible to test software for every conceivable input. The input to the software is also random. With the uncertainty associated with both the input and the software, randomness can be justified for the occurrences of errors.

5.3 SOFTWARE RELIABILITY MODELS

Reliability models can be used to predict the reliability when the software is put into operational use. Several models have been proposed [21] in the literature, and a few are described here. The software reliability models use the information of the number of errors debugged during the development of a software program. This information is used to characterize the model parameters that can then be used to predict the number of failures or some other measure of reliability in the future. The software reliability can be defined as the probability of a given software operating for a specified time period, without a software error, when used within the design limits on the appropriate machine.

5.3.1 Shooman Model

The model proposed by Shooman [17, 18] is based on the following assumptions:

1. The total number of machine language instructions in the software program is constant.
2. The number of errors at the start of integration testing is constant and decreases directly as errors are corrected. No new errors are introduced during the process of testing.
3. The difference between the errors initially present and the cumulative errors corrected represents the residual errors.
4. The failure rate is proportional to the number of residual errors.

Based on these assumptions [17],

$$e_r(x) = e(0) - e_c(x) \tag{5.1}$$

where x = debugging time since the start of system integration
$e(0)$ = errors present at $x = 0$, normalized by the total number of machine language instructions
$= E_0/I$
E_0 = number of initial errors
I = total number of machine language instructions
$e_c(x)$ = cumulative number of errors corrected by x, normalized by I
$e_r(x)$ = residual errors at x, normalized by I

Assuming failure rate to be proportional to residual errors (assumption 4),

$$\lambda_s(t) = K_s e_r(x) \tag{5.2}$$

where t = operating time of the system
K_s = constant of proportionality
$\lambda_s(t)$ = failure rate at time t

Knowing the failure rate from (5.2), the expression for reliability or survivor function [20] is

$$R(t) = \exp\left[-\int_0^t \lambda_s(x)\, dx \right]$$

$$= \exp\left[-\int_0^t K_s e_r(x)\, dx \right]. \tag{5.3}$$

Since the hazard rate is assumed independent of t in this model, this assumption amounts to a constant failure rate, and therefore,

$$\text{MTTF} = \frac{1}{\lambda_s(t)}$$

$$= \frac{1}{K_s e_r(x)} \tag{5.4}$$

Estimation of Model Parameters. By substituting for $e_r(x)$ from (5.1) into (5.4), the expression for MTTF can be written as follows:

$$\text{MTTF} = \frac{1}{K_s[e(0) - e_c(x)]}$$

$$= \frac{1}{K_s[E_0/I - e_c(x)]} \tag{5.5}$$

There are two unknowns in (5.5), K_s and E_0. These parameters can be estimated using the moment matching method [20]. Considering two debugging periods x_1 and x_2 such that $x_1 < x_2$,

$$\frac{T_1}{n_1} = \frac{1}{K_s[e(0) - e_c(x_1)]} \tag{5.6}$$

and

$$\frac{T_2}{n_2} = \frac{1}{K_s[e(0) - e_c(x_2)]} \tag{5.7}$$

where T_1, T_2 = system operating times corresponding to x_1 and x_2, respectively

n_1, n_2 = number of software errors during x_1 and x_2, respectively

From (5.6) and (5.7)

$$E_0 = \frac{I[\gamma e_c(x_1) - e_c(x_2)]}{\gamma - 1} \tag{5.8}$$

where $\gamma = \dfrac{T_1}{T_2} \cdot \dfrac{n_2}{n_1}$

$$= \frac{\text{MTTF}_1}{\text{MTTF}_2}$$

MTTF_i = the mean time to software failures corresponding to debugging time x_i.

$$= \frac{T_i}{n_i}$$

By substituting for E_0 from (5.8) into (5.6),

$$K_s = \frac{n_1}{T_1[E_0/I - e_c(x_1)]} \tag{5.9}$$

An alternative method for estimation of E_0 and K is by using maximum likelihood estimates and is discussed in reference 17.

5.3.2 The Markov Model

This model [19, 22] assumes the system to go through a sequence of "up" and "down" states. The system state is termed "up" if the first error since the start of integration and testing has not yet occurred or if the system has been restored after an error and the next error has yet to be encountered. The down state implies that an error has occurred and has not been corrected. The state transition diagram of this model is shown in Figure 5.1, in which the following notation is used:

1. State $(n-k)$ indicates that kth bug has been corrected and that $(k+1)$th error is yet to occur. This is the up state following the down state due to the kth bug.
2. State $(m-k)$ is entered when the $(k+1)$th bug is discovered. This is the down state due to $(k+1)$th error.
3. λ_k is the error occurrence rate when the system is in state $(n-k)$.
4. μ_k is the error correction rate when the system is in down state $(m-k)$.
5. $P_j(t)$ is the probability of the system being in state j at time t.

The state differential equations for this system can be easily formulated using known methods [20].

$$\dot{P}_{n-k}(t) = -\lambda_k P_{n-k}(t) + \mu_{k-1} P_{m-k+1}(t) \tag{5.10}$$

$$\dot{P}_{m-k}(t) = -\mu_k P_{m-k}(t) + \lambda_k P_{n-k}(t) \tag{5.11}$$

The initial conditions are

$$P_{m-k}(0) = 0 \qquad k = 1, 2, 3 \ldots \tag{5.12}$$

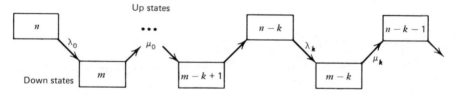

Figure 5.1 The Markov model.

and

$$P_n(0) = 1 \tag{5.13}$$

A restrictive solution of (5.10) and (5.11) assuming constant $\lambda_k = \lambda$ and constant $\mu_k = \mu$ is derived in [22]. This constraint on λ_k and μ_k is, however, easily seen to be unrealistic. A more general solution can be obtained using numerical techniques like Euler's and Runge-Kutta methods for integration. Once the state probabilities have been obtained, unavailability is calculated as [19, 22]

$$U(t) = \sum_{k=0}^{k_{max}} P_{m-k}(t) \tag{5.14}$$

The probabilities will depend on the choice of k_{max}. By choosing k_{max} large enough, $U(t)$ can be made close to the true value of $U(t)$.

5.3.3 Jelinski-Moranda Model

This model [14, 15, 21] like the Shooman model assumes an exponential probability density function for software errors. The hazard rate is assumed to be proportional to the number of remaining errors, that is,

$$\lambda_{JM}(x_i) = K_{JM}[E_0 - (i-1)] \tag{5.15}$$

where K_{JM} = constant of proportionality
x_i = time between the ith and $(i-1)$st errors discovered

The reliability function and the mean time to failure can be obtained [14] from (5.15):

$$R(t_i) = \exp[-K_{JM}(E_0 - i + 1)t_i] \tag{5.16}$$

and

$$\text{MTTF} = \int_0^\infty R(t_i)\, dt_i = \left| \frac{-1}{K_{JM}(E_0 - i + 1)} \exp[-K_{JM}(E_0 - i + 1)t_i] \right|_0^\infty$$

$$= \frac{1}{K_{JM}[E_0 - i + 1]} \tag{5.17}$$

5.3.4 Schick Wolverton Model

The Schick Wolverton model [24] assumes the hazard rate proportional to the number of remaining errors and the debugging time:

$$\lambda_{sw}(t_i) = K_{sw}[E_0 - (i-1)]x_i \tag{5.18}$$

where x_i = time interval between the $(i-1)$st and the ith error.

The reliability function is

$$R(t_i) = \exp\left[\int_0^{t_i} \lambda_{sw}(x)\, dx \right]$$

$$= \exp\frac{-K_{sw}(E_0 - i + 1)t_i^2}{2} \tag{5.19}$$

$$\text{MTTF} = \int_0^\infty R(t_i)\, dt$$

$$= \int_0^\infty \exp\left[-K_{sw}(E_0 - i + 1)t_i^2/2 \right] dt_i$$

$$= \left[\frac{\pi}{2\,K_{sw}(E_0 - i + 1)} \right]^{1/2} \tag{5.20}$$

It could possibly be argued both for and against having the hazard rate proportional to debugging time. Probably the only way to judge the suitability of this model is by fitting it to the experimental data.

5.4 MODEL VALIDATION

Four models have been described in this chapter and several more have been proposed in the literature [21]. In addition to the models described in reference 21, Bayesian models have also been proposed [10]. The true worth of a model can be measured by its ability to predict. Most of the discussion on the relative worth of the models is generally based on intuition and logical consistency. Because of the scarcity of data on software errors and lack of consistency in the available data, only a few attempts have been made for the experimental validation of these models. One such attempt has been reported in the literature [21], wherein a comparative study of the four models described in this chapter and five more models is described.

The error data used by Sukert [21] came from Software Problem Reports (SPR's) during the software development of a large command and control system. The software was written in Jovial J4 code and consisted of 249 routines with a total of 115,000 lines. Although some internal tools such as static code analyzer were used by the contractor for software development, no techniques like structured programming were used.

The data was restructured so that each entry corresponded to a single error and to delete entries due to nonsoftware errors. The data was then sorted according to the date of opening an SPR so as to provide a sequential time frame suitable for input to the models. The data on CPU time was not available from this project. A day was considered as the basic unit of debugging interval length.

Because of the unavailability of CPU data, the Shooman model could not be used. The other three models along with some modifications of these and several other models were compared and the following conclusions were drawn [21].

1. The Jelinski-Moranda and Schick-Wolverton models consistently gave higher predictions for the number of remaining errors than the actual number, that is, the prediction is conservative or pessimistic with these models.

2. For small software projects or where the testing phase is short, Jelinski-Moranda and Schick-Wolverton models appear to give a reasonable prediction for the number of remaining errors.

3. Of all the models studied, Schick-Wolverton, or a modified version of Jelinski-Moranda models appear to give the best prediction for the remaining errors for large projects or projects with a long-testing phase.

It should be remembered that even though this comparative study has produced some useful results, many more studies of this kind are needed.

5.5 SOFTWARE RELIABILITY ASSURANCE AND IMPROVEMENT

Modeling is only one aspect of software reliability and is intended to predict the number of bugs remaining in the system by using the statistical information on discovering and removing the errors. There are, however, two equally and perhaps more important areas of software reliability. These additional areas can be described as (a) designing for reliability, and (b) testing for reliability assurance. These two topics are discussed in this section.

5.5.1 Designing for Reliability

Probably the best way to have reliable software is to minimize the number and severity of bugs while a software package is being developed. There does not appear to be any proven best way of producing reliable software. There is as yet no theoretical framework for techniques for turning out error-free software. However, there appears to be an emerging consensus that certain program structuring and management techniques are conducive to developing reliable software. These techniques are usually referred to as structured programing and several techniques related to it.

Structured Programing.   Several definitions of structured programing are floating around. A more generally accepted definition of this approach appears to be coding that avoids use of GO TO statements and program design that is TOP DOWN and modular. These three features appear to enhance program reliability, readability, and maintainability.

Top Down Programing.   There are basically two ways of program design, bottom up and top down. The classical way of writing large programs is bottom up. In this approach, the program manager views the project as a whole, determines the system objectives, and then specifies the components needed for the software. The interfaces are specified and the component softwares are allocated to individual programers for development. Each programer is responsible for testing his subassembly or module before it goes into integration. The modules are integrated level by level by the most capable member of the group whose modules are being integrated. This manner of software development is similar to the one used for hardware development.

It appears, however, that an alternative approach of software development in the top down manner gives more reliable software [2]. Here the chief programer programs instead of providing supervision alone. The core of the system is written first assuming dummy subassembly at the next level. These subassemblies are developed next in likewise manner. In comparing these two approaches, an analogy with the chief surgeon is often drawn. The top down approach is like the best surgeon doing the most important or fundamental surgery himself and coordinating the less essential work performed by others [7].

GO TO Free Coding.   Dijkstra published a note in 1968 in CACM [6] entitled "Go To statement considered harmful." The title of this communication seems to have had a wide-ranging effect on contemporary programing techniques. The GO TO statement does not create errors by itself. It is the transfer of control that can create meshing of the flow of logic so that the code can become difficult to read. The avoidance of GO TO statements, on the other hand, creates more transparent and readable code. The GO TO free programs are also more straightforward to prove.

Now if it were conceded that GO TO statements should not be allowed in programing, what is the alternative? It has been shown [4] that any flow chart can be constructed using only the single entry exit structures shown in Figure 5.2. These three control structures can be used to write programs that will be free of GO TO statements and in which the program text will correspond more closely to the program execution [3]. This is best illustrated using an example. Figure 5.3 is an example of a program written using GO TO statements and the same program written using the control structures of Figure 5.2 is shown in Figure 5.4. It can be appreciated that whereas the code in Figure 5.3 jumps around the page, the one in Figure

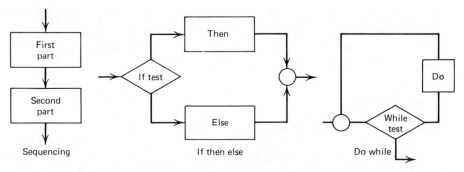

Figure 5.2   Single-entry single-exit logic structures.

A

IF B   GO TO 10

D

GO TO 30

10   E

GO TO 30

30   G   **Figure 5.3**   Program containing GO TO statements.

A

IF B   THEN   E

ELSE D

G   **Figure 5.4**   Program of Figure 5.3 without GO TO statements.

5.4 follows a sequencing process. Imagine a large piece of software written in the manner of Figure 5.3; it could be hard to follow and readily understand, whereas the code written in the manner of Figure 5.4 is more transparent. Such a code is not only readily understood but the programer is less likely to make errors.

Modular Programs.   Programers normally break down a complex software development task into separate modules. A module is used by many other modules. This also, however, increases the potential for misunderstanding and errors. It appears that this source of errors can be minimized if every module is entered only at the top and left at the bottom.

Related Ideas.   The techniques of structured programing and the concepts of modularity and top down flow have been described. These techniques have been found of much value in minimizing the number of errors in writing programs [22]. There are other related ideas like development accounting and the concept of a program librarian, which merit consideration and the reader may find references 17 and 22–23 of interest.

5.5.2 *Testing*

Even with the best programing techniques, it is likely that a piece of software will contain some bugs. The purpose of testing is to assure that the program performs according to the specification design. Test tools have been developed to assist in assessing this assurance in computer programs. These tools basically provide some numerical measure of the thoroughness with which the testing was conducted. Several such tools are available and a comparative investigation into the effectiveness of these tools is reported in reference 16.

The test tools consist of the following basic modules: (a) instrumentation module, (b) analyzer module.

The source program of the module under test is first submitted to the instrumentation module, which inserts additional statements into the module. These additional statements are called sensors and counters [16] and the process of adding these statements is called instrumentation. The functional intent of the original code must remain unchanged during the process of instrumentation, that is, the sensor and counter statements must not change the functional objectives of the program.

The instrumented package is compiled in the usual manner and the object package is executed with its test data, which results in an instrumentation data file in addition to the normal output. This instrumentation data file and the instrumented source file are then submitted to the analyzer module, which produces a report indicating the behavior of the module during execution. The following type of information is contained in such a report:

1. The number of times each statement has been executed.
2. At each branch point, how many times a particular path has been taken.
3. Time for executing each statement.

This information is useful in checking the structure of the code. It provides confidence in the logic and code of the program by ensuring that each statement and each branch path has been executed at least once. It is also possible to ensure that each subroutine has been called once. As can be inferred from the description of the testing process, these test tools can be very useful in discovering and reducing sequencing and control errors which account for approximately 20 percent of the total number of errors [16]. These structural analysis tools, however, do not test the timing and data relationships. In addition to structural testing, functional testing is also necessary if the software is time critical.

REFERENCES

1. Amster, S. J. and M. L. Shooman, "Software Reliability: An Overview," *Reliability and Fault Tree Analysis: Theoretical and Applied Aspects of System Reliability and Safety Assessment*, SIAM, Philadelphia, 1975, pp. 655–685.

2. Baker, F. T., "Chief Programmer Team Management of Production Programing," *IBM Systems J.*, **11**, 1972, pp. 56–73.

3. Benson, J. P., "Structured Programing Techniques," see Ref. 13, pp. 143–147.

4. Bohm, C. and G. Jacopini, "Flow Diagrams, Turing Machines and Languages with Only Two Formation Rules," *Commun. ACM*, **9**, 366–371 (1968).

5. Boehm, B. W., "Software and Its Impact: A Quantitative Assessment," *Datamation Mag.* **19** (5), 48–59 (May 1973).

6. Dijkstra, E. W., "GO TO Statement Considered Harmful," *Commun. ACM*, **11**, 147–148 (1968).

7. Flynn, R. J., "Design of Computer Software," *Proceedings of the 1975 Annual Reliability and Maintainability Symposium*, IEEE, New York, 1975, pp. 476–479.

8. Hecht, H., "Can Software Benefit from Hardware Experience?" *Proceedings of the 1975 Annual Reliability and Maintainability Symposium*, IEEE, New York, 1975, pp. 480–484.

9. Management Overview, IBM Installation Productivity Programs Dept., Bethesda, MD (1973).

10. Littlewood, B. and J. W. Verrall, "A Bayesian Reliability Growth Model for Computer Software," see Ref. 13, pp. 70–77.

11. Littlewood, B., "How to Measure Software Reliability and How Not to," *IEEE Trans. Reliability*, **28**, 103–110 (1979).

12. Mills, H. D., "On the Development of Large Reliable Programs," see Ref. 13, pp. 155–159.

13. MacWilliams, W. H. "Reliability of Large Real-Time Control Software Systems," Records 1973 IEEE Symposium on Computer Software Reliability, IEEE Catalog No. 73 CH 0741-9CSR, 1973, pp. 1–6.

14. Moranda, P. B. and J. Jelinski, "Software Reliability Research," In: *Statistical Computer Performance Evaluation*, Edited by Walter Freiberger, Academic, New York, 1972.

15. Moranda, P. L. and J. Jelinski, "Final Report on Software Reliability Study," McDonnell Douglas Astronautic Company, MDC Report No. 63921, Dec. 1972.

16. Reifer, D. J. and R. L. Ettenger, "Test Tools: Are They a Cure-All?" *Proceedings of the 1975 Annual Reliability and Maintainability Symposium*, IEEE, New York, 1975.

17. Shooman, M. L., "Operational Testing and Software Reliability Estimation during Program Development," *1973 IEEE Symposium on Computer Software Reliability*, IEEE, New York, 1973, pp. 51–57.

18. Shooman, M. L., "Software Reliability: Measurement and Models," *Proceedings of the 1975 Annual Reliability and Maintainability Symposium*, IEEE, New York, 1975.

19. Shooman, M. L. and A. K. Trivedi, "A Many-State Markov Model for Computer Software Performance Parameters," *IEEE Trans. Reliab.*, **R-25**, 66–68, 1976.

20. Singh, C. and R. Billinton, *System Reliability Modelling and Evaluation*, Hutchinson, London, 1977.

21. Sukert, A. N., "An Investigation of Software Reliability Models," *Proceedings of the 1977 Annual Reliability and Maintainability Symposium*, IEEE, New York, 1977.

22. Trivedi, A. K. and M. L. Shooman, "A Many State Markov Model for the Estimation and Prediction of Computer Software Performance Parameters," *Proceedings International Conference on Reliable Software*, April 21–23, 1975, Los Angeles, IEEE Cat. No. 75 CH0940-7CSR, New York.

23. USAF Report, Information Processing/Data Automation Implications of Air Force Command and Control Requirements in the 1980's, (CCIP-85), Vol. 1, SAMSO/XRS-71, April 1972.

24. Wolverton, R. W. and G. J. Schick, "Assessment of Software Reliability," TRW Systems Group, Report No. TRW-SS-72-04, Sept. 1972.

6

Mechanical Reliability

6.1 INTRODUCTION

The concept of constant failure rate is used to evaluate electronic component reliability. This concept is derived from the bathtub hazard rate belief that the failure rate remains constant during the useful life of electronic components. However, this is not normally the case when evaluating mechanical component reliability. It is an established fact in many cases that the mechanical components follow an increasing failure rate pattern that is generally represented by the exponential hazard function.

The field of mechanical reliability is relatively new as compared to the electronic reliability. The in-depth effort in this field appears to have been started since the early 1960s and may be credited to the U.S. space program. During those years, the failure of mechanical and electro-mechanical components was one of NASA's (National Aeronautics and Space Administration) prime concerns. For example, due to a mechanical failure caused by a busting high pressure tank, the SYNCOM I is believed to have been lost in space in 1963. Another typical example is the failure of Mariner III in 1964. It is also believed to have been lost due to a mechanical failure. There are several other instances where systems had mechanical failures. The researchers in the field felt that the design improvements were needed to improve reliability and longevity of mechanical and electromechanical components. Therefore, the space agency spent millions of dollars to test, replace, and redesign components such as filters, pressure switches, pressure gauges, mechanical valves, and actuators.

In 1965 NASA [80] initiated some major research projects entitled:

1. Reliability demonstration using overstress testing.
2. Reliability of structures and components subjected to random dynamic loading.
3. Designing specified reliability levels into mechanical components with time-dependent stress and strength distributions.

Ever since many publications on the subject have appeared. An up to date but selective literature on the subject is listed at the end of this chapter. In addition, a comprehensive literature survey up to 1974 on structural reliability is presented in reference 63.

At present, the most acceptable way of predicting mechanical component reliability may be by applying the interference theory. This approach is well documented in references 49 and 50. The topics presented in this chapter are as follows:

1. Statistical distributions in mechanical reliability.
2. Fundamentals of mechanical reliability.
3. Mechanical equipment basic failure modes.
4. Theory of mechanical failures.
5. Safety indices.
6. Load factors.
7. Design by reliability methodology.
8. Interference theory models.
9. Reliability optimization.

6.2   STATISTICAL DISTRIBUTIONS IN MECHANICAL RELIABILITY

This section presents failure distributions useful for representing the failure behavior of mechanical components. As compared to other distributions the extreme value distribution is the most likely candidate for the failure behavior of mechanical components. Its examples are presented in references 28 and 44.

The distributions discussed in the following sections are closely related to the reliability evaluation of mechanical components:

6.2.1   *The Exponential Distribution*

The probability density function is represented by the equation:

$$f = \lambda \exp(-\lambda t) \qquad t \geqslant 0 \qquad \lambda > 0 \qquad (6.1)$$

where t is time and λ is the constant failure rate.

The reliability function R and hazard rate z of the exponential distribution are:

$$R = \exp(-\lambda t) \qquad (6.2)$$

and

$$z = \lambda \qquad (6.3)$$

This distribution is widely used in reliability engineering. One of the reasons for its widespread use is its simplicity in performing reliability analysis. Its validity to represent a real-life failure data was first presented in reference 19.

6.2.2 The Extreme Value Distribution

The density function f of this distribution is defined by

$$f = \exp(t)\exp\{-\exp(t)\} \qquad -\infty < t < \infty \qquad (6.4)$$

where t is time. The extreme value reliability and hazard rate functions, respectively, are

$$R = \exp\{-\exp(t)\} \qquad (6.5)$$

and

$$z = \exp(t) \qquad (6.6)$$

This distribution was first used to analyze flood data by Gumbel [28]. Therefore, it is sometimes known as the Gumbel's distribution. The failure behavior of many mechanical components may be represented by this distribution. From more fundamental considerations, this distribution can be developed by considering a corrosion process [66].

6.2.3 The Weibull Distribution

The Weibull density function is given by

$$f = \beta\lambda t^{\beta-1}e^{-\lambda t^{\beta}} \qquad \text{for} \quad \beta>0 \quad \lambda>0 \quad t\geqslant 0 \qquad (6.7)$$

where λ = the scale parameter
β = the shape parameter
t = time

Weibull reliability and hazard functions are

$$R = e^{-\lambda t^{\beta}} \qquad (6.8)$$

and

$$z = \beta\lambda t^{\beta-1} \qquad (6.9)$$

This distribution was developed by Weibull [99], who described some of its applications. Ball bearing failures applications are given in reference [64].

The exponential ($\beta = 1$) and Raleigh ($\beta = 2$) are the special cases of this distribution.

6.2.4   The Mixed Weibull Distribution

This distribution was first presented by Kao [43]. He applied it to measure reliability of electron tubes. The probability density function is defined as

$$f = \frac{k\alpha_1}{\beta_1}\left(\frac{t}{\beta_1}\right)^{\alpha_1 - 1} \exp\left(-\frac{t}{\beta_1}\right)^{\alpha_1}$$

$$+ \frac{(1-k)}{\beta_2}\alpha_2\left(\frac{t-\theta}{\beta_2}\right)^{\alpha_2 - 1} \exp\left\{-\left(\frac{t-\theta}{\beta_2}\right)^{\alpha_2 - 1}\right\} \qquad (6.10)$$

for $\beta_1, \beta_2 > 0, 0 < \alpha_1 < 1, \alpha_2 > 1, \theta > 0, 0 \leqslant k \leqslant 1$

The reliability expression for the above density function is

$$R = 1 - k\left[1 - \exp\left(\frac{t}{\beta_1}\right)^{\alpha_1}\right] - (1-k)\left[1 - \exp\left\{-\left(\frac{t-\theta}{\beta_2}\right)^{\alpha_2}\right\}\right] \quad (6.11)$$

6.2.5   The Gamma Distribution

The gamma probability density function is defined as

$$f = \frac{\lambda^\beta t^{\beta - 1} \exp(-\lambda t)}{\Gamma(\beta)} \qquad \text{for} \quad \lambda > 0 \quad \beta > 0 \quad t \geqslant 0 \qquad (6.12)$$

where $\Gamma(\beta) = \displaystyle\int_0^\infty t^{\beta - 1} e^{-t}\, dt$

$\beta = $ the shape parameter
$\lambda = $ the scale parameter

The reliability and hazard rate expressions are

$$R = \left[\int_t^\infty x^{\beta - 1} \exp(-\lambda x)\, dx\right]\lambda^\beta / \Gamma(\beta) \qquad (6.13)$$

and

$$z = \frac{t^{\beta - 1}\exp(-\lambda t)}{\displaystyle\int_t^\infty x^{\beta - 1}\exp(-\lambda x)\, dx} \qquad (6.14)$$

This distribution is an extended version of the exponential distribution. It was applied to the life test problems by Gupta and Groll [26].

The gamma distribution is related to the exponential and Chi-squared distributions. For its applications one should consult reference [57].

6.2.6 The Log-Normal Distribution

The probability density function is

$$f = \frac{1}{(t-\theta)\sqrt{2\pi}\ \sigma}\exp-\frac{\{\ln(t-\theta)-\mu\}^2}{2\sigma^2} \tag{6.15}$$

for $t>\theta>0$

where $\mu =$ is the mean
$\sigma =$ the standard deviation

The reliability and hazard rate expressions for the above function are given by

$$R = \frac{1}{\sqrt{2\pi}\ \sigma}\int_t^\infty \frac{1}{(t-\theta)}e^{-\{\ln(t-\theta)-\mu\}^2/2\sigma^2}\,dt$$

for $t>\theta$ $\tag{6.16}$

and

$$z = \frac{[1/(t-\theta)]e^{-\{\ln(t-\theta)-\mu\}^2/2\sigma^2}}{\int_t^\infty [1/(t-\theta)]e^{-\{\ln(t-\theta)-\mu\}^2/2\sigma^2}\,dx} \tag{6.17}$$

Normally, the hazard rate of this distribution is an increasing function of time followed by a decreasing function. The hazard rate approaches zero for initial and infinite times. A representative example of this distribution is the failures due to fatigue cracks.

6.2.7 The Fatigue Life Distribution Models

These distribution models were presented by Birnbaum and Saunders [9], who proposed two-parameter distributions. The main applications of a family of distributions are to characterize failures due to fatigue.

The probability density function is defined as

$$f = \frac{(t^2-\lambda^2)}{2\sqrt{2\pi}\ \alpha^2\lambda t^2(t/\lambda)^{1/2}-(\lambda/t)^{1/2}}\exp\left[-\frac{1}{2\alpha^2}\left(\frac{t}{\lambda}+\frac{\lambda}{t}-2\right)\right]$$

for $t>0$ $\alpha,\lambda>0$

where α and λ are the shape and scale parameters, respectively. Readers requiring in depth material on these distributions should consult references 27, 44, and 91. Other hazard rate models are presented in references [67, 91].

6.3 FUNDAMENTALS OF MECHANICAL RELIABILITY

Like any other field of reliability engineering, mechanical reliability is also a joint responsibility of design and reliability engineers. A reliability engineer augments the designer's knowledge with design review procedures and statistical analysis; however, the designer still remains the key person to ensure component or system reliability.

The old concept of merely good design practices is not satisfactory to ensure reliability of a complex system. Reference 72 lists several reasons for the discipline of mechanical reliability.

1. *Lack of design experience.* Changes in technology are quite rapid and the mechanical designers no longer have the time to master the design especially when a complex equipment is designed for use in aerospace or military applications.
2. *Cost and time constraints.* Because of the cost and time involved, the designer cannot learn from past mistakes. In other words the cut-and-try approach cannot be used.
3. *Optimization of resources.* The workable design is no longer considered sufficient. The design must be optimized subject to constraints such as reliability, cost, weight, performance, and size.
4. *Stringent requirements and severe environments.* Because of large-scale investments in developing systems to be used under severe environments (military and space) the reliability problem becomes important.
5. *Influence from electronic reliability.* The vastly improved techniques for predicting electronic component reliability also stimulated similar developments in mechanical engineering.

6.4 MECHANICAL EQUIPMENT BASIC FAILURE MODES

Unlike electronic components, the mechanical components have numerous failure modes. Some of the basic failure modes pertaining to mechanical equipment are fatigue, leakage, wear, thermal shock, creep, impact, corrosion, erosion, lubrication failure, elastic deformation, surface fatigue, radiation damage, spalling, corrosion wear, delamination, and buckling. These basic failure modes are described in detail in reference 72. Some of these

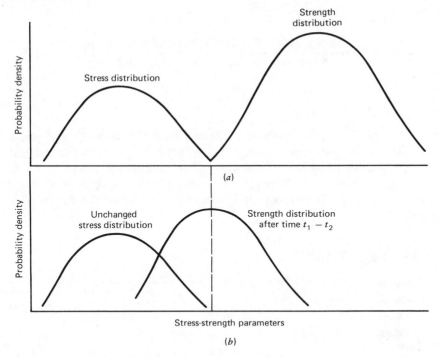

Figure 6.2 (*a*) Stress-strength distribution at time t_1; (*b*) stress-strength distribution at time t_2.

6.6 SAFETY INDICES

The safety factor approach is a conventional design technique. This method uses safety margins and safety factors that are simply arbitrary multipliers. In some cases, these factors provide satisfactory design, if they are established from the past experience. In the days of modern technology, however, the new design involves new applications and new materials and more consistent methods are needed. The mechanical component design based entirely upon safety factors, could be misleading, and may be costly due to overdesign or could end up in a catastrophic failure due to underdesign. It is emphasized that whenever a designer makes use of safety factors these must be based upon considerable experience on similar items.

6.6.1 Safety Factor

There are several different ways of defining a safety factor as outlined in reference 55. In reference 10, the theoretical definition of a safety factor, is defined as

$$s_f = \frac{\text{average value of failure governing strength, } \mu_s}{\text{average value of failure governing stress, } \mu_{ss}}$$

$$= \frac{\mu_s}{\mu_{ss}} \geqslant 1 \tag{6.19}$$

This is a good measure particularly when both the strength and stress distributions are normally distributed. This factor in a mechanical design is always equal to or greater than unity. The concept of safety factor is illustrated in Figure 6.3. When the variation of stress and/or of strength is large, the safety factor becomes meaningless because the failure rate is positive.

6.6.2   Safety Margin

In reference 55 the safety margin is defined in the following ways:

$$s_m = s_f - 1 \tag{6.20}$$

or

$$s_m = \frac{\mu_s - \mu_{\text{max}}}{\sigma_s} \tag{6.21}$$

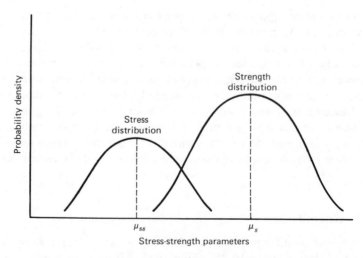

Figure 6.3   Safety factor for stress-strength distribution.

where μ_s = average strength

μ_{max} = maximum stress

σ_s = standard deviation of strength

and

$$\mu_{max} = \mu_{ss} + k\sigma_{ss} \qquad (6.22)$$

where μ_{ss} = mean stress

σ_{ss} = standard deviation of stress

Normally, the value of k is between 3 and 6. It can be observed from discussion on safety margins that (a) it is a random variable just like its counterpart, the safety factor, and (b) it presents the idea of separation of stress and strength mean values.

Example 1. Suppose

$$\sigma_{ss} = 200 \text{ psi}, \; k = 4, \; \sigma_s = 900 \text{ psi}$$

$$\mu_s = 25{,}000 \text{ psi} \quad \text{and} \quad \mu_{ss} = 12{,}000 \text{ psi}$$

Find the safety margin for given data. By substituting the above information in (6.21), we get

$$s_m = \frac{25{,}000 - (12{,}000 + 4 \times 200)}{900}$$

$$= \frac{25{,}000 - 12{,}800}{900} = \frac{122}{9}$$

$$\simeq 13.6$$

6.7 LOAD FACTORS

In the last decade or so it has been realized that the loads as well as the capacities of structures are not necessarily deterministic but are probabilistic, because of the random variation in magnitude and the random occurrence of loads. In this section we discuss the determination of load factors in the structural design. In references 78 and 85 this subject is discussed in detail. In this section we mainly deal with the dead and live loads. Earlier analysis on the topic were initiated by the authors of references 78 and 85.

6.7.1 Deterministic Resistance with Normally Distributed Loads

For the deterministic resistance the design load, D_1, may be formulated as follows:

$$D_1 = L_d \mu_d + L_1 \mu_1 \tag{6.23}$$

where L_d = the dead load factor
  L_1 = the live load factor
  μ_d = nominal mean dead load
  μ_1 = nominal mean live load

Suppose the live and dead loads are normally distributed random variables with mean values of μ_1 and μ_d, respectively. The design load then also follows the normal law. In the case of independent dead and live loads, the design load, D_1, may be described by (6.24).

$$D_1 = (\mu_d + \mu_1) + c\sqrt{\sigma_d^2 + \sigma_1^2} \tag{6.24}$$

where σ_d, σ_l are the standard deviations of the dead and live load and c is the reliability coefficient for the combined dead and live load. Also, the design load in terms of component loads may be written as

$$D_1 = m_d + m_1 = (\mu_d + c'\sigma_d) + (\mu_1 + c'\sigma_1) \tag{6.25}$$

where c' is the reliability coefficient for each load component, that is,

$$c' = \frac{c}{\sigma_1 + \sigma_d} \sqrt{\sigma_1^2 + \sigma_d^2}$$

  m_d = magnitude of component dead load

  m_1 = magnitude of component live load

By manipulating (6.23), (6.24), (6.25), we obtain the following load factor equations:

$$L_d = \frac{m_d}{\mu_d} = 1 + scV_d \tag{6.26}$$

$$L_1 = \frac{m_1}{\mu_1} = 1 + scV_1 \tag{6.27}$$

where $s = \dfrac{(\sigma_d^2 + \sigma_1^2)^{1/2}}{\sigma_d + \sigma_1}$

$$V_d = \frac{\mu_d}{\sigma_d} \qquad V_1 = \frac{\mu_1}{\sigma_1}$$

where V_d is the coefficient of variation of dead load and V_1 is the coefficient of variation of live load.

6.7.2 Normally Distributed Loads and Resistance

When the resistance follows a normal distribution, (6.26) and (6.27) are modified to the following form:

$$L_d = \frac{1 + sc V_d}{1 - c V_R} \tag{6.28}$$

$$L_1 = \frac{1 + sc V_1}{1 - c V_R} \tag{6.29}$$

where V_R is the resistance coefficient of variation. When V_R is equal to zero, the resistance follows the deterministic law. The value of c can be determined from (6.30) when loads and resistance follow the normal distribution:

$$c = \frac{\left[(V_R \mu_R)^2 + \sigma_1^2 + \sigma_d^2 \right]^{1/2}}{V_R \mu_R + \left(\sigma_1^2 + \sigma_d^2 \right)^{1/2}} \cdot c^* \tag{6.30}$$

where μ_R is the mean resistance and c^* is the reliability coefficient of the system.

By substituting (6.30) into (6.28) and (6.29), we can determine the load factors for any desired level of reliability. Therefore the value of the c^* can be obtained from the table of the error function. For example at the desired level of reliability, say $R = 0.9901$, $c^* = 2.33$. For a solved numerical example see reference 86.

6.8 "DESIGN BY RELIABILITY" METHODOLOGY

The "design by reliability" methodology is described in considerable detail in references 50 and 49. To design an equipment or a component by taking reliability into consideration, the following steps are needed:

1. Define the design problem in question.
2. List and identify all the associated design variables and parameters in the problem.
3. Perform failure modes, effect, and criticality analysis (FMECA).
4. Determine the failure governing stress and strength functions and distributions of a failure mode.
5. Use failure governing stress and strength distribution to evaluate each critical failure mode reliability.

6. Iterate the design until the assigned reliability goals are met.
7. Optimize design under specified constraints such as cost, weight, volume, reliability, maintainability, safety performance, and so on.
8. Repeat the above steps for each vital component or device of a system.
9. Calculate the system reliability by applying the classical reliability theory.
10. Iterate the design until the specified system reliability goal is fulfilled.

Step 4 is probed in depth in the following section:

6.8.1 Determination of Failure Governing Stress Distribution

The following steps are to be followed to determine the failure governing stress distribution:

1. List and identify all the important failure modes.
2. In the case of a fracture failure mode, if any, determine the most likely locations where the combination of stresses are likely to act which may result in component failure.
3. At each location calculate the nominal stress of components.
4. Evaluate maximum value of each component stress with the use of necessary stress modifying factors.
5. At each location combine all the stresses into the failure governing stress in accordance with particular failure mode being considered.
6. In the failure governing stress equation determine each nominal stress, modifying factor and parameter distribution.
7. Determine a failure governing stress distribution from the step 6 distributions.
8. Repeat steps 2–7 for each significant failure mode listed in step 1.

Readers who require more information should consult references 51 and 54.

6.8.2 Determination of the Failure Governing Strength Distribution

To determine failure governing strength distribution, the following steps are outlined:

1. Set up the failure governing strength procedure by taking the failure modes into consideration. This criterion should be based upon the one used to determine failure governing stress.
2. Evaluate the nominal strength.

3. Use appropriate strength factors to modify nominal strength. This is to convert the nominal strength obtained under the standardized and idealized test conditions.
4. Determine the nominal strength distribution, of each modifying factor and parameter associated with the failure governing strength equation.
5. Establish the failure governing strength distribution by utilizing the normal distributions of step 4.

For more detailed information regarding the determination of the failure governing strength distribution, the interested reader should consult references 49 and 50.

6.9 RELIABILITY DETERMINATION—CONSTANT STRESS-STRENGTH INTERFERENCE THEORY MODELS

This section deals with situations in which the stress and strength are represented by well-defined probability density functions. Furthermore, the stress-strength distributions are not time dependent.

When the probability density functions of both stress and strength are known, the component reliability may be determined analytically. Reliability is defined as the probability that the failure governing stress will not exceed the failure governing strength. In a mathematical equation it can be written as

$$R=P(s<S)=P(S>s) \tag{6.31}$$

where R = the reliability of a component or a device
  P = the probability
  S = the strength
  s = the stress

Equation (6.31) can be rewritten in the following form:

$$R=\int_{-\infty}^{\infty} f_{st}(s)\left[\int_{s}^{\infty} f_{Sth}(S)\,dS\right]ds \tag{6.32}$$

where $f_{st}(s)$ = the probability density function of the stress, s
  $f_{Sth}(S)$ = the probability density function of the strength S

Reliability for a single failure mode can also be computed from (6.33) on the basis that the stress will be less than the strength:

$$R=\int_{-\infty}^{\infty} f_{Sth}(S)\left[\int_{-\infty}^{S} f_{st}(s)\,ds\right]dS \tag{6.33}$$

The above equation may be used to obtain numerical solutions if the analytical solution is difficult to obtain. In addition, when the empirical data is sufficient but the stress or strength distribution cannot be identified, the graphical approach can be applied to obtain component reliability.

6.9.1 Reliability Calculation by Graphical Approach

This technique makes use of the Mellin transforms, which can be applied to any distribution. The Mellin transforms of the reliability equation (6.32) are defined as

$$M = \int_s^\infty f_{Sth}(S)\, dS$$

$$= 1 - F_{Sth}(s) \tag{6.34}$$

and

$$L = \int_0^s f_{st}(s)\, ds = F_{st}(s) \tag{6.35}$$

Equation 6.35 may be rewritten as

$$dL = f_{st}(s)\, ds \tag{6.36}$$

By substituting (6.36) and (6.34) into (6.32) we get

$$R = \int_0^1 M\, dL \tag{6.37}$$

Obviously, L takes values from 0 to 1. Therefore, if we plot (6.37), that is, M versus L, the area under the curve will represent the single failure mode component reliability. A typical plot of (6.37) is shown in Figure 6.4. Simpson's rule can be used to calculate area under the M versus L curve.

Example 2. Suppose the strength of a component follows the Rayleigh distribution with known scale parameter value of 15,000 psi. Similarly, the stress follows a Weibull distribution with the shape parameter equal to 3 and the scale parameter value of 12,000 psi.

Therefore, the stress and strength density functions become

$$f_{Sth}(S) = \frac{2}{15,000}\left(\frac{S}{15,000}\right)\exp\left[-\left(\frac{S}{15,000}\right)^2\right] \tag{6.38}$$

and

$$f_{st}(s) = \frac{3}{12,000}\left(\frac{s}{12,000}\right)^2\exp\left[-\left(\frac{s}{12,000}\right)^3\right] \tag{6.39}$$

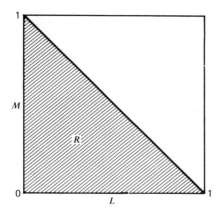

Figure 6.4 A hypothetical plot of L versus M, where the shaded area represents the component reliability, R.

By substituting (6.38) and (6.39) into (6.34) and (6.35), respectively, we get:

$$M = 1 - F_{Sth}(s) = \exp\left[-\left(\frac{s}{15,000}\right)^2\right] \qquad (6.40)$$

$$L = F_{st}(s) = 1 - \exp\left[-\left(\frac{s}{12,000}\right)^3\right] \qquad (6.41)$$

Table 6.1 presents tabulation for M and L for the various values of s. Figure 6.5 shows a plot of values for M and L from Table 6.1. Using

Table 6.1

s	M	L
0	1	0
2,000	0.98	0.005
4,000	0.93	0.04
6,000	0.85	0.12
8,000	0.75	0.26
10,000	0.64	0.44
12,000	0.53	0.63
14,000	0.42	0.8
16,000	0.32	0.91
18,000	0.24	0.97
20,000	0.17	0.99
22,000	0.12	$0.997 \simeq 1$

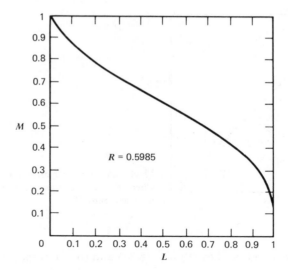

Figure 6.5 *M* versus *L* plot.

Simpson's rule, the component reliability *R* is estimated from Figure 6.5

$$R = \frac{0.25}{3} \{y_0 + 4y_1 + 2y_2 + 4y_3 + y_4\}$$

$$R = \frac{0.25}{3} \{1 + 4 \times 0.75 + 2 \times 0.61 + 4 \times 0.46 + 0.12\}$$

$$R = 0.5985$$

Reliability calculation when stress and strength data can not be represented by any existing distribution is discussed in reference 79.

6.9.2 Analytical: Constant Stress-Strength Interference Theory Models*

This section presents three interference theory models when probability density functions are defined.

Component Reliability Determination for Exponential Stress and Strength Distributions. Both stress and strength probability functions are defined as

$$f_{st}(s) = \lambda_{st} e^{-\lambda_{st} s} \qquad 0 \leqslant s < \infty \qquad (6.42)$$

and

$$f_{Sth}(S) = \lambda_{Sth} e^{-\lambda_{Sth} S} \qquad 0 \leqslant S < \infty \qquad (6.43)$$

---

*For these models it is assumed that the component has only one significant failure mode.

where $\lambda_{st}=$ the reciprocal of the mean value of stress, $\bar{s}$
  $\lambda_{Sth}=$ the reciprocal of the mean value of strength, $\bar{S}$

By utilizing expression (6.33), the component reliability, R_c, can be determined:

$$R_c=\int_0^\infty f_{Sth}(S)\left[\int_0^S f_{st}(s)\,ds\right]dS \qquad (6.44)$$

where

$$\int_0^S f_{st}(s)\,ds=\int_0^S \lambda_{st}e^{-\lambda_{st}s}\,ds=1-e^{-\lambda_{st}S} \qquad (6.45)$$

Therefore by substituting (6.45) into (6.44) we get

$$R_c=\int_0^\infty \lambda_{Sth}e^{-\lambda_{Sth}S}\left[1-e^{-\lambda_{st}S}\right]dS$$

$$=1-\int_0^\infty \lambda_{Sth}e^{-(\lambda_{st}+\lambda_{Sth})S}\,dS$$

$$=1-\frac{\lambda_{Sth}}{\lambda_{Sth}+\lambda_{st}}\int_0^\infty (\lambda_{st}+\lambda_{Sth})e^{-(\lambda_{st}+\lambda_{Sth})S}\,dS$$

$$R_c=\frac{\lambda_{st}}{\lambda_{st}+\lambda_{Sth}} \qquad (6.46)$$

Dividing numerator and denominator of expression (6.46) by λ_{st} we get:

$$R_c=\frac{1}{1+\lambda_{Sth}/\lambda_{st}}\quad\text{for }\bar{S}\neq0$$

$$=\frac{1}{1+\rho} \qquad (6.47)$$

where $\rho=\bar{s}/\bar{S}$ for $\bar{S}\geqslant\bar{s}$, $\rho\leqslant1$. Values of R_c are presented in Table 6.2 for the various values of ρ. A plot of (6.47) is shown in Figure 6.6.

6.9.3   Component Reliability Determination when Stress and Strength Follow Rayleigh Distribution

Both Rayleigh stress and strength density functions are defined as follows:

$$f_{st}(s)=2k_{st}se^{-k_{st}s^2}\qquad 0\leqslant s<\infty \qquad (6.48)$$

and

$$f_{Sth}(S)=2k_{Sth}Se^{-k_{Sth}S^2}\qquad 0\leqslant S<\infty \qquad (6.49)$$

Table 6.2

ρ	$1+\rho$	$R_c = \dfrac{1}{1+\rho}$
1	2	0.5
0.9	1.9	0.53
0.8	1.8	0.56
0.7	1.7	0.59
0.6	1.6	0.63
0.5	1.5	0.67
0.4	1.4	0.71
0.3	1.3	0.77
0.2	1.2	0.83
0.1	1.1	0.91
0	1	1

where $k_{st}=$ the stress parameter
$k_{Sth}=$ the strength parameter

Component reliability is determined by substituting (6.48) and (6.49) into (6.32):

$$R_c = \int_0^{\infty} 2k_{st}se^{-k_{st}s^2}2\left[\int_s^{\infty}k_{Sth}Se^{-k_{Sth}S^2}\,dS\right]ds$$

$$= \int_0^{\infty} 2k_{st}se^{-k_{st}s^2}2\left[e^{-k_{Sth}s^2}\right]ds = 2\int_0^{\infty}2k_{st}se^{-(k_{st}+k_{Sth})s^2}\,ds$$

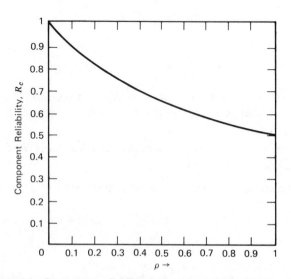

Figure 6.6 Component reliability versus mean stress-strength ratio.

Let

$$a = (k_{st} + k_{Sth})$$

$$= 4k_{st}\int_0^\infty s e^{-as^2}\, ds$$

$$\therefore R_c = 4k_{st}\frac{\Gamma(1/2)}{2a} = \frac{2k_{st}\sqrt{\pi}}{k_{st} + k_{Sth}} \tag{6.50}$$

6.9.4 Component Reliability Calculation with Normally Distributed Stress and Gamma Distributed Strength

Stress and strength probability density functions are defined as

$$f_{st}(s) = \frac{1}{\sigma_{st}\sqrt{2\pi}} e^{-(s-\mu_{st})^2/2\sigma_{st}} \qquad 0 \leqslant s \leqslant \infty \tag{6.51}$$

and

$$f_{Sth}(S) = \frac{1}{\Gamma(\beta)}\lambda^\beta S^{\beta-1} e^{-\lambda S} \qquad 0 \leqslant S \leqslant \infty \tag{6.52}$$

where β and λ are the shape and scale parameters, respectively, and μ_{st} and σ_{st} are the mean and the standard deviation, respectively. By substituting the probability density functions (6.51) and (6.58) into (6.32) and integrating, the following reliability expression is obtained,

$$R_c = \sum_{\theta=0}^{\theta=\beta-1} \sum_{s=0}^{s=\theta} \frac{(\lambda\sigma_{st})^\theta}{\theta!} e^{-\frac{\lambda}{2}(2\mu_{st}-\sigma_{st}^2\lambda)} XYZ \tag{6.53}$$

where

$$X = \left[2^{\theta/2-1}\binom{\theta}{s} a^{\theta-c/\sqrt{\pi}} \right]$$

$$Y = \Gamma\left(\frac{s+1}{2}\right)$$

$$Z = \left[1 - I\left(r, \frac{\theta-1}{2}\right) \right]$$

where I is the incomplete gamma function.

$$r = \frac{1}{\sqrt{(s+1)}}\frac{1}{\sqrt{2}}\left(\frac{\mu_{st}-\sigma_{st}^2\lambda}{\sigma_{st}}\right)^2$$

For the detailed derivation of (6.53) see reference 102. Many other inter-ference theory models to calculate component reliability are developed in references 45, 83, 101, and 102. These models are developed for the following:

1. Normally distributed stress and strength.
2. Log-normally distributed stress and strength.
3. Exponentially (normally) distributed strength and normally (exponen-tially) distributed stress.
4. Gamma distributed stress and strength.
5. Weibull distributed strength and normally distributed stress.
6. Weibull distributed stress and strength.
7. Weibull distributed strength and extreme value distributed stress.
8. Maxwellian distributed stress and Weibull distributed strength.

6.9.5 Component Reliability with Multiple Failure Modes

Reliability of a component with many independent failure modes is given by

$$R = \prod_{i=1}^{n} R_i \tag{6.54}$$

where R = the overall component reliability
n = the number of significant failure modes
R_i = the reliability of a significant failure mode i

Similarly, the system reliability can be computed for a series configura-tion, the component reliability being obtained by applying (6.54) or di-rectly from the stress-strength models (i.e., if the component under study has only one significant failure mode).

6.9.6 Chain Model

This model represents a situation in which a chain is composed of n number of identical series links subject to the same environmental stress [83]. The probability of any link having strength S_0 or greater is given by

$$P(S_{Sth} \geqslant S_0) = \int_{S_0}^{\infty} f_{Sth}(S) \, dS \tag{6.55}$$

In the case of n number of identical and independent links, the probability that the chain has strength S_0 or greater is given by

$$P(S_{Sth} \geqslant S_0) = \left[\int_{S_0}^{\infty} f_{Sth}(S) \, dS \right]^n \tag{6.56}$$

To obtain the probability density function of the chain strength, $f_{cSth}(S)$, differentiate expression (6.56) with respect to S:

$$f_{cSth}(S) = n \left[\int_{S_0}^{\infty} f_{Sth}(S) \, dS \right]^{n-1} f_{Sth}(S) \tag{6.57}$$

When all the chain links are under the same environmental stress, the chain reliability R_{ch} can be obtained by substituting (6.57) into (6.33):

$$R_{ch} = \int_0^{\infty} \left\{ \left[\int_0^S f_{st}(s) \, ds \right] n \left[\int_S^{\infty} f_{Sth}(S) \, dS \right]^{n-1} f_{Sth}(S) \right\} dS \tag{6.58}$$

Reliability of the above equation may be determined by graphical, analytical or numerical technique.

6.9.7 Stress-Strength Time-Dependent Models

In the previous sections, we considered stress-strength models where stress and strength were independent of time. In real life, however, this may not be necessarily true. The component strength may change with time and a component may experience repeated application of stresses. In other words, the stress or load may follow a random pattern with respect to time t. A hypothetical pattern is shown in Figure 6.7.

This area of mechanical reliability still remains to be explored further. The interested readers are advised to consult references 10, 45, 84, and 87.

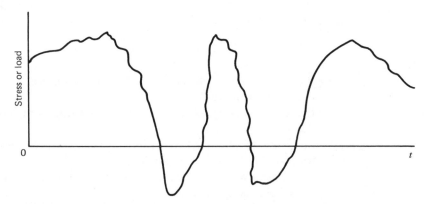

Figure 6.7 A hypothetical random stress spectrum.

6.10 OPTIMIZATION OF MECHANICAL COMPONENT RELIABILITY

A redundant system can be optimized subject to constraints such as cost, weight, and volume. To optimize system reliability, traditional operations research techniques such as Lagrange multiplier, linear, integer, and dynamic programming are applicable. These techniques can be used to optimize reliability of mechanical components, also.

6.10.1 *Reliability Optimization of a Mechanical Component with Normally Distributed Stress and Strength*

The following reliability equation is taken from reference 45:

$$R = \int_{-n}^{\infty} \frac{1}{\sqrt{2\pi}} e^{-y^2/2} \, dy \qquad (6.59)$$

where $\qquad n = (\bar{S} - \bar{s})(\sigma_{Sth}^2 + \sigma_{st}^2)^{-1/2}$
$\bar{s} =$ mean stress
$\bar{S} =$ mean strength
$\sigma_{Sth}, \sigma_{st} =$ standard deviations of strength and stress

It is assumed that to formulate this model, the stress and strength are statistically independent. To maximize component reliability, it is obvious that the value of lower limit of expression (6.59) should be as low as possible. Therefore, the equation to minimize total cost subject to desired component reliability may be formulated as follows:

$$\text{minimize } k = k_1(\bar{S}) + k_2(\sigma_{Sth}) + k_3(\bar{s}) + k_4(\sigma_{st})$$

$$\text{subject to } (\bar{S} - \bar{s})(\sigma_{Sth}^2 + \sigma_{st}^2)^{-1/2} \geqslant y \qquad (6.60)$$

where $\qquad k =$ the total cost
$k_1(\bar{S}) =$ the cost function of the mean strength (monotonically increasing function)
$k_3(\bar{s}) =$ the cost function of the mean stress (monotonically decreasing function)
$k_2(\sigma_{Sth}) =$ the strength standard deviation cost function (monotonically decreasing function)
$k_4(\sigma_{st}) =$ the stress standard deviation cost function (monotonically decreasing function)
$y =$ obtained by the coupling equation for the desired reliability level

The Lagrangian equation for the above problem becomes:

$$F\left(\bar{S}, \bar{s}, \sigma_{Sth}, \sigma_{st}, \lambda\right) = k + \lambda\left[\bar{S} - \bar{s} - y\left(\sigma_{Sth}^2 + \sigma_{st}^2\right)^{1/2}\right] \qquad (6.61)$$

To find optimum solution, differentiate (6.61) with respect to each variable $\lambda, \bar{S}, \bar{s}, \sigma_{Sth}, \sigma_{st}$, and equate each differentiation to zero. The following equations were obtained:

$$\bar{S} - \bar{s} - y\left(\sigma_{Sth}^2 + \sigma_{st}^2\right)^{1/2} = 0 \qquad (6.62)$$

$$\dot{k}_4(\sigma_{st}) = \lambda y \sigma_{st} \Big/ \sqrt{\sigma_{Sth}^2 + \sigma_{st}^2} \qquad (6.63)$$

$$\ddot{k}_2(\sigma_{Sth}) = \lambda y \sigma_{Sth} \Big/ \sqrt{\sigma_{Sth}^2 + \sigma_{st}^2} \qquad (6.64)$$

$$k_3'(\bar{s}) = \lambda \qquad (6.65)$$

$$k_1''(\bar{S}) = -\lambda \qquad (6.66)$$

where single overdots and primes represent partial derivative with respect to $\sigma_{st}, \bar{s}$, respectively, and double overdots and primes represent partial derivative with respect to $\sigma_{Sth}, \bar{S}$, respectively. The value of $\bar{S}, \bar{s}, \sigma_{Sth}, \sigma_{st}$, and λ can be found by solving (6.62)–(6.66) to obtain all local optima. To choose a global optimal solution, determine the objective function (6.60) for all the local optimal solutions. For a more detailed analysis and examples on the mechanical component reliability optimization, one should consult references 45 and 95.

6.11 CONCLUDING REMARKS

Although the interference stress-strength modeling is a promising technique for calculating the reliability of a mechanical component, there are several problem areas to be overcome. Some of these problems are outlined as follows:

1. The representative stress and environmental condition under which the component will operate may be difficult to estimate with certainty at the design stage because of the lack of field data.
2. Most of the material properties are time dependent. For some practical purposes this factor may be disregarded because of their slow change, but generally, the time dependency can not be ignored. Due to the lack of variability data of material properties, further assumptions regarding time dependency may be required.
3. Although there is no lack of mathematical techniques or the probabilistic models for the reliability evaluation, further refinement to these

techniques and models will be useful to improve the mechanical reliability prediction.

REFERENCES

1. Ang, A. H. S., "Extended Reliability Basis of Structural Design Under Uncertainties," *1970 Annals of Reliability and Maintainability Conference*, pp. 642–649. IEEE, New York.

2. Aitchison, J. and J. A. C. Brown, *The Lognormal Distribution*, Cambridge University Press, New York, 1963.

3. Austin, W. H., "Development of Improved Gust Load Criteria for USAF Aircraft," *1967 Annals of Reliability and Maintainability Conference*, pp. 68–74. IEEE, New York.

4. Bazovsky, I., "Reliability of Rifles, Machine Guns and Other Small Arms," *Proceedings Annual Symposium on Reliability*, 1971, pp. 71–73. IEEE, New York.

5. Bell, R. and R. Mioduski, "Extension of Life of US Army Trucks," *1976 Annual Reliability and Maintainability Symposium*, pp. 200–205. IEEE, New York.

6. Bennet, S. B., A. L. Ross, and P. Z. Zemanick, Editors, *Failure Prevention and Reliability*, Society of Mechanical Engineers, New York, 1977.

7. Bhattacharyya, G. K. and R. A. Johnson, *Stress-Strength Models for System Reliability*, *Reliability and Fault Tree Analysis*, SIAM, Philadelphia, 1975, pp. 509–532.

8. Billet, R. B., "Reliability of Hydraulic Controls in Space Vehicles," *National Symposium on Reliability*, 1964, pp. 69–84. IEEE, New York.

9. Birnbaum, Z. W. and S. C. Saunders, "A New Family of Life Distributions," *J. Appl. Probl.*, pp. 319–327 (Oct. 1969).

10. Bompass-Smith, J. H., *Mechanical Survival: The Use of Reliability Data*, McGraw-Hill, London, 1973.

11. Bompass-Smith, J. H., "The Determination of Distributions that Describe the Failures of Mechanical Components," *1969 Annals of Reliability and Maintainability Conference*, pp. 343–356. IEEE, New York.

12. Bratt, M. J., H. A. Truscott, and G. W. Weber, "Probabilistic Strength Mapping: Reliability Vs Life Prediction Tool," *1968 Annals of Reliability and Maintainability Conference*, pp. 501–510. IEEE, New York.

13. Bratt, M. J., G. Reethoff, and G. W. Wieber, "A Model for Time Varying and Interfering Stress/Strength Probability Density Distributions with Consideration for Failure Incidence and Property Degradation," *Proceedings 3rd Annual Aerospace Reliability and Maintainability Conference*, 1969, pp. 566–575. IEEE, New York.

14. Brewer, J. W., "Interim Scale Reliability Statements Consistent with Conventional Materials Strength Criteria," *1971 Annals of Reliability and Maintainability Conference*, pp. 63–71. IEEE, New York.

15. Burns, J. J., "Reliability of Nuclear Mechanical Systems," *1975 Annual Reliability and Maintainability Symposium*, pp. 163–169. IEEE, New York.

16. Carter, A. D. S., *Mechanical Reliability*, Wiley, London, 1972.

17. Chester, L. B., C. F. Nolf, and D. Kececioglu, "Combined Bending-Torsion Fatigue Reliability—III," *1975 Annual Reliability and Maintainability Symposium*, pp. 511–518. IEEE, New York.

18. Collins, J. A., G. T. Hagan, and H. M. Bratt, "Helicopter Failure Modes and Corrective Actions," *1975 Annual Reliability and Maintainability Symposium*, pp. 504–510. IEEE, New York.

19. Davis, D. J., "An Analysis of Some Failure Data," *J. Amer. Stat. Assoc.*, pp. 113–150 (1952).

20. Dehardt, J. H. and H. D. McLaughlin, "Using Bayesian Methods to Select a Design with Knoron Reliability Without a Confidence Coefficient," *1966 Annals of Reliability and Maintainability Conference*, IEEE, New York, 1966, pp. 611–617.

21. Dennis, N. G., "PMR, NDE, Design Practices; Present and Future," *1977 Annual Reliability and Maintainability Symposium*, Philadelphia, pp. 164–170. IEEE, New York.

22. Dillin, A. L., "Reliability Assessment of Army Weapons and Weapon Systems," *Proceedings of the Annual Symposium on Reliability*, IEEE, New York, 1971, pp. 74–76.

23. Disney, R. L. and N. J. Sheth, "The Determination of the Probability of Failure by Stress/Strength Interference Theory," *1968 Annual Symposium on Reliability*, IEEE, New York, 1968, pp. 417–422.

24. Forrestor, E. R. and V. H. Thevenow, "Designing for Expected Fatigue Life," *1968 Annals of Reliability and Maintainability*, pp. 511–519. IEEE, New York.

25. Ghane, P. M., "Quality and Safety Factors in Reliability," *1970 Annals of Reliability and Maintainability*, IEEE, New York, 1970, pp. 637–641.

26. Gupta, S. and P. Groll, "Gamma Distribution in Acceptance Sampling Based on Life Tests," *J. Amer. Stat. Assoc.*, Dec. 1961, pp. 942–970. IEEE, New York.

27. Gross, A. J. and V. A. Clark, *Survival Distribution*: *Reliability Applications in the Biomedical Sciences*, Wiley, New York, 1975.

28. Gumbel, E. J., *Statistics of Extremes*, Columbia University Press, New York, 1958.

29. Hald, A., *Statistical Theory with Engineering Applications*, Wiley, New York, 1952.

30. Haugen, E. B., "Statistical Methods for Structural Reliability Analysis," *National Symposium on Reliability*, IEEE, New York, 1964, pp. 97–121.

31. Haviland, R. P., *Engineering Reliability and Longlife Design*, Van Nostrand, Princeton, 1964.

32. Heller, R. A. and H. S. Heller, "Analysis of Early Failures in Unequal," *1973 Annual Reliability and Maintainability Symposium*, IEEE, New York, 1973, pp. 198–200.

33. Heller, R. A. and M. Shinozuka, "State-of-the-Art: Reliability Techniques in Materials and Structures," *1970 Annals of Reliability and Maintainability Conference*, IEEE, New York, 1970, pp. 635–636.

34. Ingram, G. E., C. R. Herrmann, and E. L. Welker, "Designing for Reliability Based on Probabilistic Modeling Using Remote Access Computer Systems," *1968 Annals of Reliability and Maintainability Conference*, IEEE, New York, 1968, pp. 492–500.

35. Ingram, G. E., E. B. Haugen, C. Dicks, and S. Wilson, "Panel Discussion—Structural Reliability," *National Symposium on Reliability*, IEEE, New York, 1965, pp. 154–169.

36. Inoue, K. and H. Daito, "Safety Analysis of Automobile Brake Systems by Fault Tree Methods," *Proceedings HOPE International JSME Symposium*, Japan Society of Mechanical Engineers, Tokyo, 1977, pp. 213–220.

37. Japan Society of Mechanical Engineers, *Proceedings of the HOPE International Symposium*, 1979.

38. Johnson, C. W. and R. E. Maxwell, "Reliability Analysis of Structures—A New Approach," *1976 Annual Reliability and Maintainability Symposium*, pp. 213–217. IEEE, New York.

39. Johnson, W. S., R. A. Heller, and J. N. Yang, "Flight Inspection Data and Crack Initiation Times," *1977 Annual Reliability and Maintainability Symposium*, Philadelphia, pp. 148–154. IEEE, New York.

40. Jones, L. G. and D. Thompson, "Essential Elements of Analysis From Army Test and Field Data," *Proceedings Annual Symposium on Reliability*, 1971, pp. 85–90. IEEE, New York.

41. Kalivoda, F. E. and K. W. Yun, "Modeling Mechanical System Accelerated Life Tests," *1976 Annual Reliability and Maintainability Symposium*, pp. 206–212. IEEE, New York.

42. Kao, J. H. K., "A Summary of Some New Techniques for Failure Analysis," *Proceedings Sixth National Symposium on Reliability*, Washington D.C., 1960, p. 191. IEEE, New York.

43. Kao, J. H. K., "A Graphical Estimation of Mixed Weibull Parameters in Life-Testing of Electron Tubes," *Technometrics*, 1, pp. 389–407, 1959.

44. Kao, J. H. K., "Statistical Models in Mechanical Reliability," *National Symposium on Reliability*, IEEE, New York, 1965, pp. 240–247.

45. Kapur, K. C. and L. R. Lamberson, *Reliability in Engineering Design*, John Wiley, New York, 1977.

46. Karnopp, D., "Structural Reliability Predictions Using Finite Element Programs," *1971 Annals of Reliability and Maintainability Conference*, IEEE, New York, 1971, pp. 72–80.

47. Kececioglu, D. and A. Koharcheck, "Wear Reliability of Aircraft Splines," *1977 Annual Reliability and Maintainability Symposium*, Philadelphia, IEEE, New York, 1977, pp. 155–163.

48. Kececioglu, D., "Why Design By Reliability?" *1968 Annals of Reliability and Maintainability Conference*, IEEE, New York, 1968, p. 491.

49. Kececioglu, D., "Reliability Analysis of Mechanical Components and Systems," *Nucl. Eng. Design*, 19, pp. 259–290 (1972).

50. Kececioglu, D., "Fundamentals of Mechanical Reliability Theory and Applications to Vibroacoustic Failures," *Proceedings of Reliability Design for Vibroacoustic Environments*, ASME, New York, 1974, pp. 1–38.

51. Kececioglu, D., J. W. McKinley, and M. Saroni, "A Probabilistic Method of Designing a Specified Reliability into Mechanical Components with Time Dependent Stress and Strength Distributions," The University of Arizona, Tucson, Arizona, Jan. 1967. (NASA Report under Contract NGR 03-002-044.)

52. Kececioglu, D., "Probabilistic Design Methods for Reliability and their Data and Research Requirements," *Failure Prevention and Reliability Conference Proceedings*, ASME, New York, 1977.

53. Kececioglu, D., L. B. Chester, and T. M. Dodge, "Alternating Bending—Steady Torque Fatigue Reliability," *1974 Annual Reliability and Maintainability Symposium*, IEEE, New York, 1974, pp. 153–173.

54. Kececioglu, D. and D. Cormier, "Designing a Specified Reliability Directly into a Component," *Proceedings 3rd Annual Conference on Aerospace Reliability and Maintainability*, 1964, Washington D.C., IEEE, New York, 1964, pp. 546–565.

55. Kececioglu, D. and E. B. Haugen, "A Unified Look at Design Safety Factors, Safety Margins and Measures of Reliability," *1968 Annals of Reliability and Maintainability Conference*, IEEE, New York, 1968, pp. 520–530.

56. Kececioglu, D. B., R. E. Smith, and E. A. Felsted, "Distributions of Strength in Simple Fatigue and the Associated Reliabilities," *1970 Annals of Reliability and Maintainability Conference*, IEEE, New York, 1970, pp. 659–672.

57. King, J. R., *Probability Charts for Decision Making*, Industrial Press, New York, 1971.

58. Kirkpatrick, I., "Predicting Reliability of Electromechanical Devices," *Sixth National Symposium on Reliability and Quality Control*, IEEE, New York, 1960, pp. 272–281.

59. Konno, K., K. Nakano, and Y. Yoshimura, "A New Accelerated Fatigue Test, The Effective Random Peak Method," *1975 Annual Reliability and Maintainability Symposium*, IEEE, New York, 1975, pp. 263–268.

60. Kullman, L. W. and G. W. Phillips, "Reliability Engineering Disciplines Applied to Commercial Weapon Systems—Guns and Ammunition," *National Symposium on Reliability*, IEEE, New York, 1964, pp. 122–137.

61. Kurtz, P. H., "Reliability Study of a Hydraulic Control System · Using the Hybrid Computer," *1971 Annals of Reliability and Maintainability Conference*, IEEE, New York, 1971, pp. 81–85.

62. Lambert, R. G., "Mechanical Reliability for Low Cycle Fatigue," *1978 Annual Reliability and Maintainability Symposium,*, IEEE, New York, 1978, pp. 179–183.

63. Lemon, G. H. and S. D. Manning, "Literature Survey on Structural Reliability," *IEEE Trans. Reliab.*, **R-23**, (October 1974).

64. Lieblein, J. and M. Zelen, "Statistical Investigation of the Fatigue Life of Deep-Groove Ball Bearings," *J. Res. Nat. Bur. Stand.*, 5 (Res. Paper 2719), 273–316 (1956).

65. Liebowitz, H., "Navy Reliability Research," *1967 Annals of Reliability and Maintainability Conference*, IEEE, New York, 1967, pp. 33–53.

66. Lloyds, D. and M. Lipow, *Reliability: Management, Methods and Mathematics*, Prentice-Hall, Englewood Cliffs, NJ, 1961.

67. Mann, N. R., R. E. Shafer, and N. D. Singpurwalla, *Methods for Statistical Analysis of Reliability and Life Data*, Wiley, New York, 1974.

68. Manning, S. D. and G. H. Lemon, "Plan for Developing Structural Criteria for Composite Airframes," *1974 Annual Reliability and Maintainability Symposium*, IEEE, New York, 1974, pp. 155–162.

69. Marble, Q. G., "Improving Mechanical Reliability of Digital Computers," *National Symposium on Reliability and Quality Control*, IEEE, New York, 1965, pp. 136–143.

70. Martin, P., "Reliability in Mechanical Design and Production," *Proceedings Generic Techniques in Systems Reliability Assessment*, Noordhoff-Leyden, Amsterdam, 1976, pp. 267–271.

71. Matney, V. D., "Reliability, Pollutants and Aluminum Raw," *1974 Annual Reliability and Maintainability Symposium*, IEEE, New York, 1974, pp. 174–178.

72. *Mechanical Reliability Concepts*, ASME Design Engineering Conference, ASME, New York, 1965.

73. Mesloh, R., "Reliability Design Criteria for Mechanical Creep," *1966 Annals of Reliability and Maintainability Conference*, IEEE, New York, 1966, pp. 590–597.

74. Moreno, F. J., "Reliability Estimate of a Space Deployable Antenna," *1973 Annual Reliability and Maintainability Symposium*, IEEE, New York, 1973, pp. 182–185.

75. Naresky, J. J., "Reliability and Maintainability Research in the USAF," *1966 Reliability and Maintainability Conference*, IEEE, New York, 1966, pp. 769–787.

76. Nilsson, S. O., "Reliability Data on Automotive Components," *1975 Annual Reliability and Maintainability Symposium*, IEEE, New York, 1975, pp. 276–279.

77. Niyogi, P. K., "Application of Statistical Methods and Information Theory to Structural Reliability Estimates," PhD Dissertation, Department of Civil Engineering, University of Pennsylvania, Philadelphia, PA, 1968.

78. Niyogi, P. K., H. C. Shah, K. D. Doshi, and W. Tang, *Statistical Evaluation of Load Factors for Concrete Bridge Design*, Chicago, April 1969.

79. *Quality Assurance: Reliability Handbook*, Headquarters, US Army Material Command, October 1968, (AMCP 702-C, AD 702 936), NTIS, Springfield, VA.

80. Redler, W. M., "Mechanical Reliability Research in the NASA," *1966 Annals of Reliability and Maintainability Conference*, IEEE, New York, 1966, pp. 763–768.

81. Reethof, G., "Session Organizer's Report," *1973 Annual Reliability and Maintainability Symposium*, IEEE, New York, 1973, p. 181.

82. Reethof, G., "State-of-the-Art—Mechanical and Structural Reliability," *1971 Annals of Reliability and Maintainability Conference*, IEEE, New York, 1971, p. 62.

83. Roberts, N., *Mathematical Methods in Reliability Engineering*, McGraw-Hill, New York, 1964.

84. Schatz, R., M. L. Shooman, and L. Shaw, "Application of Time Dependent Stress-Strength Models of Non-Electrical and Electrical Systems," *Proceedings of the Reliability and Maintainability Symposium*, January 1974, IEEE, New York, pp. 540–547.

85. Shah, H. C., "Use of Maximum Entropy in Estimating the Damage Distribution of a Single Degree of Freedom System Subjected to Random Loading," *1966 Annals of Reliability and Maintainability Conference*, IEEE, New York, 1966, pp. 598–604.

86. Shah, H. C. and W. H. Tang, "Statistical Evaluation of Load Factors in Structural Design," *1970 Annal of Reliability and Maintainability Conference*, IEEE, New York, 1970, pp. 650–658.

87. Shaw, L., M. Shooman, and R. Schatz, "Time-Dependent Stress-Strength Models for Non-Electrical and Electronic Systems," *1973 Annual Reliability and Maintainability Symposium*, IEEE, New York, 1973, pp. 186–197.

88. Shinozuka, M. and M. Hanai, "Structural Reliability of A Simple Rigid Frame," *1967 Annals of Reliability and Maintainability Conference*, IEEE, New York, 1967, pp. 63–67.

89. Shinozuka, M. and H. Itagaki, "On the Reliability of Redundant Structures," *1966 Annals of Reliability and Maintainability Conference*, IEEE, New York, 1966, pp. 605–610.

90. Shooman, M. L., *Probabilistic Reliability: An Engineering Approach*, McGraw-Hill, New York, 1968.

91. Singpurwella, N. D., "Statistical Fatigue Models: A Survey," *IEEE Trans. Reliab.*, **R-20**, pp. 185–189 (1971).

92. Spoormaker, J. L., "Reliability Prediction of Hairpin Type Springs," *1977 Annual Reliability and Maintainability Symposium*, Philadelphia, IEEE, New York, 1977, pp. 142–147.

93. Spoormaker, J. L., "Design of Reliable Plastic Assemblies," *1975 Annual Reliability and Maintainability Symposium*, IEEE, New York, pp. 498–503.

94. Stiles, E. M., "Reliability in Mass-Produced Consumer Products," *National Symposium on Reliability*, IEEE, New York, 1964, pp. 85–96.

95. Taraman, S. I. and K. C. Kapur, "Optimization Considerations in Design Reliability by Stress-Strength Interference Theory," *IEEE Trans. Reliability*, **24**, (1975), pp. 136–138.

96. Thomas, J. M., S. Hanagnd, and J. D. Hawk, "Decision Theory in Structural Reliability," *1975 Annual Reliability and Maintainability Symposium*, IEEE, New York, pp. 255–262.

97. Tumolillo, T. A., "Methods for Calculating the Reliability Function for Systems Subjected to Random Stresses," *IEEE Trans. Reliability*, **23**, 256–262 (1974).

98. Weaver, L. and T. Scarlett, "Reliability and Failure Distributions of Inertial Sensors," *National Symposium on Reliability and Quality Control*, IEEE, New York, 1965, pp. 144–153.

99. Weibull, W., "A Statistical Distribution Function of Wide Applicability," *J. Appl. Mech.*, **18**, 293–297 (1951).

100. Welker, D. R. and H. N. Buchanan, "Safety-Availability Study Methods Applied to BART," *1975 Annual Reliability and Maintainability Symposium*, IEEE, New York, 1975, pp. 269–275.

101. Yadav, R. P. S., "Component Reliability Under Environmental Stress," *Microelectron. Reliability*, **13**, pp. 473–475 (1974).

102. Yadav, R. P. S., "A Reliability Model for Stress Vs. Strength Problem," *Microelectron. Reliability*, **12**, pp. 119–123 (1973).

103. Yao, J. T. P. and H. Y. Yeh, "Safety Analysis of Statically Indeterminate Trusses," *1967 Annals of Reliability and Maintainability Conference*, IEEE, New York, 1967, pp. 54–62.

7

Human Reliability

7.1 INTRODUCTION

Numerous systems are interconnected by human links. In the earlier reliability analysis, attention was directed only to equipment, and reliability of the human element was neglected.

Williams [94] recognized this shortcoming in the late 1950s and pointed out that realistic system reliability analysis must include the human aspect. Ever since the beginning of the last decade there has been a considerable interest in human-initiated equipment failures and their effect on system reliability.

According to reference 50 about 20–30 percent of failures, directly or indirectly are due to human error. Furthermore, according to reference 19 about 10–15 percent of the total failures are directly related to human errors. These are mainly due to wrong actions, maintenance errors, misinterpretation of instruments, and so on.

Subsequent work by others is listed in Section 7.10. This research deals mainly with the human error data banks, human error classification schemes, determining the significance of errors to system operation, human error allocation, and human reliability modeling in continuous time domain.

7.1.1 Human Reliability Definition

According to reference 49, human reliability is defined as the probability that a job or task will be successfully completed by personnel at any required stage in system operation within a required minimum time (if the time requirement exists).

7.1.2 Human Error

Human error is defined [19] as a failure to perform a prescribed task (or the performance of a prohibited action), which could result in damage to equipment and property or disruption of scheduled operations. In real life most systems require some human participation irrespective of the degree of automation. It is said that wherever people are involved, errors will be

161

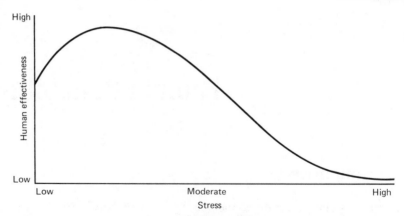

Figure 7.1   A hypothetical human effectiveness-stress curve.

made. These errors occur regardless of their training, skill, or experience. Therefore, predicting equipment reliability without considering human reliability will not present a true picture of that reliability.

7.2   HUMAN STRESS-PERFORMANCE EFFECTIVENESS

According to reference 19, the human performance and stress follow the relationship shown in Figure 7.1. This relationship shows that the human error rate for a particular task follows a curvelinear relation to the imposed stress. At a very low stress, the task is dull and unchallenging; therefore most operators will not perform effectively and the performance will not be at the optimal level. When the stress is at a moderate level, the operator performs at his optimum level. The moderate level may be interpreted as high enough stress to keep the operator alert. At a still higher stress level, the human performance begins to decline. This decline is mainly due to fear, worry, or other types of psychological stress. It follows from Figure 7.1 that at the highest stress level, the human reliability is at its lowest level.

7.3   CONCEPT OF HUMAN ERROR

According to reference 33, a human error occurs if any one of the following happens:

1. The operator or any human pursues a wrong goal.
2. The required goal is not met because the operator acted wrongly.
3. The operator fails to act in the moment of need.

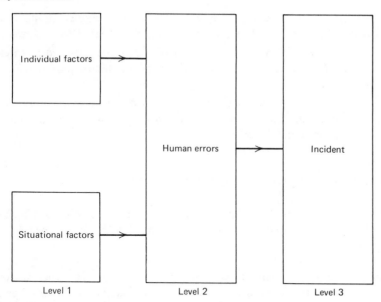

Figure 7.2 Levels of human error.

The human errors may be divided into three levels as shown in Figure 7.2. The situation may be corrected at each level of human error, shown in Figure 7.2. For example future human errors may be prevented at level 1. At level 2 a future incident can be avoided by correcting the wrong action due to human error. In the case of level 3 one could prevent the same situation from occurring again.

7.4 TYPES OF HUMAN ERROR

The author of reference 50 has categorized the human errors as follows:

1. *Design error*. This error results from inadequate design. For example, the controls and displays are so far apart that an operator finds difficulty in using both of them effectively.
2. *Operator error*. This occurs if the operating personnel fail to follow correct procedures, or there is lack of correct procedures.
3. *Fabrication error*. This error occurs at the fabrication stage due to (a) poor workmanship, for example, incorrect soldering; (b) use of wrong material; (c) the fabrication is not according to the blueprint requirement.
4. *Maintenance error*. This type of error occurs in the field. It is normally due to incorrect installation or repair of the equipment.

5. *Contributory error.* Under this category we can represent those errors that are difficult to define either as human or related to equipment.
6. *Inspection error.* This error is associated with accepting out-of-tolerance component or equipment; or rejecting in-tolerance equipment or component.
7. *Handling error.* The handling error occurs due to inappropriate storage or transport facilities, which are not in accordance with the manufacturer's recommendations.

7.5   CAUSES OF HUMAN ERRORS

This section presents the main causes of human errors some of which have already been discussed in Section 7.4. Some of the main causes are as follows:

1. Poor training or skill of the operating personnel. In other words the operators or maintenance staff are not adequately equipped to perform the prescribed task.
2. Inadequate maintenance or operating procedures for the operating personnel.
3. Poor job environments, for example, accessibility, crowded space, and temperature.
4. Poor or inadequate handling of equipment or tools.
5. Poor motivation for the operators or the maintenance personnel which effects their performance from being at optimum level.

7.6   HUMAN UNRELIABILITY DATA BANKS

The material presented in this section is taken from reference 47. Therefore, the interested reader can consult this reference for further details. This paper presents a brief review of existing methods to develop human reliability data banks. The major emphasis of this publication is upon the estimation data collected from expert opinions. The author states that there is a lack of human data compared to the techniques available to predict human reliability.

The human error data banks may be divided into the following three categories:

1. Experimentally based data banks.
2. Field-based data banks.
3. Subjectively based data banks.

7.6.1 Experimentally Based Data Banks

This type of data bank is based upon laboratory sources and is gathered in the laboratory. The main advantage of this data is that it is the least influenced by the subjective elements that may produce some error. Therefore, one can have more confidence in such data banks. One must, however, be aware that no matter how carefully these data banks are developed, there is always a considerable amount of subjective element present.

The well-known data bank based on the experimental findings is the *data store* [52]. This data bank is based upon 164 selected studies.

7.6.2 Field-Based Data Banks

These data banks are based upon the operational data and are more realistic than the experimentally based data banks. However, the field-based data banks are rather difficult to establish because these banks are based upon real activities occurring in the operating environment. The results obtained from these banks are more satisfactory than those obtained from the experimental sources whose tasks are often contrived.

At present there are two noteworthy field-based data banks, which are described in references 93 and 78. The one presented in reference 93 is called the Operational Performance Recording and Evaluating Data System (OPREDS), which permits the automatic monitoring of all operator actions. However, it is only applicable to limited cases (e.g., switch actions). The other proposed data bank is called the Sandia Human Error Rate Bank (SHERB) [78].

7.6.3 Subjectively-Based Data Banks

These data banks are based upon expert opinions and have two attractive features:

1. They are comparatively easy to develop because a large amount of data can be collected from a small number of expert respondents.
2. They are cheaper to develop.

The subjective-based data is obtained by using less rigorous techniques such as DELPHI [13]. This technique narrows the guess-estimate variations of the field experts by feeding back the end result of the study to individual judges or experts. It makes them reconsider their guess-estimates until some form of consensus is arrived. This method is already effective at the Naval Personnel Research and Development Center [36].

The following requirements must be satisfied if these banks are to be used in the human reliability analysis:

1. *Validity.* A subjective data bank will contain some error. Therefore, we should be prepared to accept a somewhat lower accuracy of such data banks as compared to the experimental data ones.
2. *Expert Judgement.* The subjective data should be collected only from those personnel who are recognized as highly skilled to perform tasks in question and in addition, have observed others performing such tasks. For example, it is better to obtain data from operators rather than the human reliability experts.
3. *Performance Dimensions.* The technique to be used should be decided very carefully, keeping in mind the dimensions of the performance being estimated.
4. *Judgment Description Level.* The performance-shaping factors associated with these estimates must be determined at an early stage. Furthermore, the types of errors to be included for a particular task should be clarified.
5. *Procedure Specification.* To obtain subjective estimates, the applicable procedure should be specified, for example, whether it is DELPHI or paired comparisons.

The main advantage of this type of data bank is the coverage of a wide range of parameters for which failure data is required.

7.7 HUMAN RELIABILITY MODELING IN CONTINUOUS TIME DOMAIN

The material presented in this section is based on reference 63. Some of the typical examples of such tasks are scope monitoring, aircraft maneuvering, and missile countdown. This type of modeling is analogous to the classical reliability modeling.

The generalized human performance reliability function for continuous time tasks is derived in the following section. (Note: for discrete case consult reference 62.)

7.7.1 Human Performance Reliability Function in Continuous Time Domain

Although all human tasks are not in continuous time domain, tasks such as vigilance, monitoring, and tracking fall in this category. In the case of continuous tasks, the probability of occurrence of human error in the time interval, (δt given E_1) is given by

$$P(E_2/E_1) = e(t)\,\delta t \qquad (7.1)$$

where $e(t)$ = the human error rate at time t; this is analogous to the hazard rate, $z(t)$, in the classical reliability theory
E_1 = an errorless performance event of duration t
E_2 = an event that the human error will occur in time interval $(t, t+\delta t)$

The joint probability of the errorless performance may be expressed as follows:

$$P(\bar{E}_2/E_1)p(E_1) = P(E_1) - P(E_2/E_1)P(E_1) \qquad (7.2)$$

where $\bar{E}_2$ denotes the event that error will not occur in interval $[t, t+\delta t]$. The above equation may be rewritten as

$$R_h(t) - R_h(t)P(E_2/E_1) = R_h(t+\delta t) \qquad (7.3)$$

where $R_h(t)$ is human reliability. Expression 7.2 represents an errorless performance probability over intervals $[0, t]$ and $[t, t+\delta t]$.
By substituting (7.1) into (7.3) we get

$$\frac{R_h(t+\delta t) - R_h(t)}{\delta t} = -e(t)R_h(t) \qquad (7.4)$$

In the limiting case, the above expression becomes

$$\frac{dR_h(t)}{dt} = -e(t)R_h(t) \qquad (7.5)$$

To solve the differential equation we may write for known initial conditions

$$\int_0^t e(t)\,dt = -\int_1^{R_h(t)} \frac{1}{R_h(t)} dR_h(t) \qquad (7.6)$$

The solution to the differential equation (7.5) is

$$R_h(t) = e^{-\int_0^t e(t)\,dt} \qquad (7.7)$$

This is the general expression to compute human reliability.

7.7.2   Reliability Quantifiers for Time Continuous Human Performance Tasks

These parameters are analogous to the classical reliability theory. Time continuous human performance task quantifiers are defined as follows:

Mean Time to Human Initiated Failure (MTHIF).   This index is analogous to the mean time to failure (MTTF) in the classical reliability theory. This quantifier is used for the time continuous tasks such as undershooting a landing aircraft or overpressurizing a missile fuel tank.

Mean-Time-to-First-Human-Error (MTFHR).   This quantifier is analogous to the mean time to first failure (MTTFF) in the classical theory. The MTFHR may be used for cases where the occurrence of the first human error is highly critical.

Mean Time Between Human Errors (MTBHE).   This is known as the mean time between human errors. It is directly translated from the mean time between failure (MTBF) as known in the classical reliability theory. This indicator may be used where the human errors are not so critical. For example, it may be used for measuring the occurrence of defective parts due to human errors at a production line.

7.7.3   Experimental Justification of the Time Continuous Human Performance Model

To justify time continuous task model discussed earlier, the authors of reference 63 have developed a simple model to obtain human error data. The main feature of this experiment was to observe a clock-type light display. The operator was required to respond to a failed light event by pressing a hand held switch.

The following types of data was collected from this experiment:

1. *Miss error.* The operator (subject) did not detect the failed light.
2. *False alarm error.* The operator (subject) responds in such a way as if a failed-light event has occurred when it did not occur in reality.

The failure data collected from this study was analyzed by graphical technique and the Kolmogorov-Smirnov *d* statistic.

This study reported that the human error rate is a time variant. Furthermore, this experiment tested the following types of errors:

1. Times to first miss error.
2. Times to false alarm error.
3. Combined miss and false alarm error.

The Weibull, gamma, and log-normal density functions emerged as the representative distributions for the goodness of fit.

7.7.4 Human Performance Effectiveness Function (Correctability) in Time Continuous Domain

The correctability function $C_h(t)$ concerns with the correction of the self-generated human errors. In reference 63, it is defined as the probability that a task error will be corrected in time t subject to stress constraint inherent in the nature of the task and its environment. In other words, the correctability function may be defined as

$$C_h(t) = P \{\text{correction of error in time } t/\text{stress}\} \qquad (7.8)$$

The time derivative of not-correctability function $\overline{C}_h(t)$ may be defined as

$$\overline{C}'_h(t) = -\frac{1}{N} N'_{\overline{C}}(t) \qquad (7.9)$$

where the prime denotes differentiation with respect to time t. N is the total number of times task correction accomplished after time t. $N_{\overline{C}}(t)$ is the number of times task not completed after time t.

Equation 7.9 may be rewritten in the following form:

$$N\{N_{\overline{C}}(t)\}^{-1}\overline{C}'_h(t) = N'_{\overline{C}}(t)\{N_{\overline{C}}(t)\}^{-1} \qquad (7.10)$$

The right-hand side of (7.10) represents instantaneous task correction rate $C_R(t)$. Hence, (7.10) may be rewritten as

$$\{\overline{C}_h(t)\}^{-1}\overline{C}'_h(t) + C_R(t) = 0 \qquad (7.11)$$

By solving the above differential equation for given initial conditions we get

$$\overline{C}_h(t) = e^{-\int_0^t C_R(t)\, dt} \qquad (7.12)$$

since

$$C_h(t) + \overline{C}_h(t) = 1$$

Therefore,

$$C_h(t) = 1 - e^{-\int_0^t C_R(t)\, dt} \qquad (7.13)$$

The above equation is a general expression. It holds for both constant and instantaneous correction rates. The experimental results with data, for the above function are presented in reference 63. This experiment dealt with the operation of a standard *E*-type manual control stick grip, subject to two degrees of freedom representing the pitch and roll motions of an aircraft in response to the instrument altitude pointer movement.

These results indicate that for both vigilance and compensatory tracking tasks, the Weibull density function is a suitable fit for the time to first error correction. On the other hand, the log-normal is equally applicable for the time for correction of errors.

7.8   HUMAN ERROR PREDICTION TECHNIQUE

This technique is relatively well known among the human reliability experts. It is known as THERP (technique for human error rate prediction). THERP, which is discussed in detail in reference 79, is based upon the classical analysis method. The basic steps associated with THERP are

1. List main system failure events.
2. List and analyze human related functions.
3. Obtain estimates for the human error rates.
4. Determine human error effects on the system failure events in question.
5. Make necessary recommendations and necessary changes in the system in question. At the end compute new failure rates for the system under study.

7.8.1   Probability Tree Analysis

This is one of the main techniques for human reliability analysis. Success or failure of each critical human action or associated event is assigned a conditional probability. The outcome of each event is represented by the branching limbs of the probability tree. The total probability of success for a particular operation is obtained by summing up the associated probabilities with the end point of the success path through the probability tree diagram. This technique, with some refinement, can include factors such as time stress, emotional stress, interaction stress, interaction effects, and equipment failures.

Some of the advantages of this technique are as follows:

1. It serves as a visibility tool.
2. The mathematical computations are simplified, which in turn decrease the probability of occurrence of errors due to computation.
3. The human reliability analyst can estimate conditional probability readily, which may otherwise be obtained from the complicated probability equations.

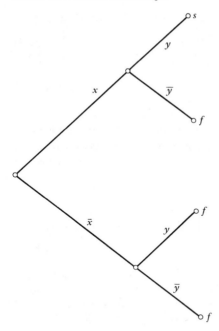

Figure 7.3 A hypothetical task probability tree diagram.

Example. Assume that an operator performs two tasks, say *x* and *y* (the task *x* is performed before *y*). In addition assume that tasks *x* and *y* can be performed either correctly or incorrectly. In other words the incorrectly performed tasks are the only errors that can occur in this situation. Draw the probability tree for this example and obtain the overall system probability to perform incorrect task. In this example we assume that the probabilities are statistically independent.

This example states that the operator can perform task *x* correctly or incorrectly. Later, the operator may proceed to perform task *y* which also has two different possibilities (correct and incorrect). The following notations were used to define the probability tree diagram as shown in Figure 7.3:

P_s = probability of task success
P_f = probability of failure to accomplish required task
s = success
f = failure
P_x = probability of success in performing task *x*
P_y = probability of success in performing task *y*
$P_{\bar{x}}$ = probability of failure to perform task *x*
$P_{\bar{y}}$ = probability of failure to perform task *y*

The probability of success, P_s, can be written from Figure 7.3 as follows:

$$P_s = P_x \cdot P_y \tag{7.14}$$

Similarly, the failure probability, P_f, can be written directly from Figure 7.3 as follows:

$$P_f = P_x P_{\bar{y}} + P_{\bar{x}} P_y + P_{\bar{x}} P_{\bar{y}} \qquad (7.15)$$

$$= 1 - P_x P_y \qquad (7.16)$$

It can be noticed from Figure 7.3 that the only way the system task can be performed successfully is that both the tasks x and y must be done correctly. Therefore the probability of performing system task correctly is simply given by $P_x P_y$. This technique is described in more detail in reference 79.

7.9 HUMAN RELIABILITY ANALYSIS APPLIED TO NUCLEAR PLANTS

There is no single technique that can readily be applicable to the nuclear power plants. The technique such as THERP may be applied to predict human reliability. However, the following performance-shaping factors [77] are to be considered in the human reliability analysis when applied to the nuclear power plant.

1. Training and practice quality.
2. Quality and existence of written instructions as well as the method of proper usage.
3. Quality of human engineering as applied to the nuclear power plant controls and displays.
4. Type of the display feedback. For example, there may be too many displays competing for the operator attention.
5. Human action independence.
6. Redundancy concerning humans.
7. Psychological stress.

Once these shaping factors have been considered, one should proceed to estimate the human error rate. Human error rate estimates then should be included in the Fault Tree Analysis. This type of analysis is probed in depth in reference 61.

REFERENCES

1. Altman, J. W., "Classification of Human Errors," In: Edited by W. B. Askren, *Proceedings Symposium Reliability of Human Performance in Work*, Wright-Patterson AFB, Ohio, Rep AMRL-TR-67-88, 1967.

2. Bailey, R. W., S. T. Demers, and A. I. Lebowitz, "Human Reliability in Computer-Based Business Information Systems," *IEEE Trans. Reliability*, **R-22**, pp. 140–147 (Aug. 1973).

3. Baron, S., "Application of the Optimal Control Model for the Human Operator to Reliability Assessment," *IEEE Trans. Reliab.*, **22**, 157–164 (August 1973).

4. Barone, M. A., "A Methodology to Analyse and Evaluate Critical Human Performance," *Annals of Reliability and Maintainability Conference*, IEEE, New York, 1966.

5. Birdsall, C. R., "Comments on Psychological Reliability, Man-Machine Systems," *IEEE Trans. Reliab.*, November 1971. pp. 260–261.

6. Blanchard, R. E., "Survey of Navy User Needs for Human Reliability Models and Data Report No. 102-1," Naval Underwater Systems Center, New London Laboratory, New London, CT, December 1972.

7. Brown, E. S., "System Safety and Human Factors: Some Necessary Relationships," *Proceedings of the Annual Reliability and Maintainability Symposium*, 1974.

8. Celinski, O. and M. Master, "An Activity Model for Predicting the Reliability of Human Performance," *Annual Reliability and Maintainability Symposium*, 1974.

9. Cooper, J. I., "Human-Initiated Failures and Malfunction Reporting," *IRE Trans. Human Factors Electron.*, **HFE**, 104–109 (September 1961).

10. Cornog, D. Y. and A. H. Ruder, "A Postal Service Field Evaluation of Letter Sorting," *Proceedings of the Annual Reliability and Maintainability Symposium*, 1977.

11. Crawford, B. M., "The Human Component in Systems," *Annals of Reliability and Maintainability*, 1971.

12. Cunningham, C. E. and W. Cox, *Human Factors in Maintainability*, *Applied Maintainability Engineering*, Wiley-Interscience, New York, 1972.

13. Dalkey, N. and F. Helmer, "An Experimental Application of the DELPHI Method to the Use of Experts," *Manag. Sci.*, **9**, 458–467, (1963).

14. De Callies, R. N., "Human Reliability in the Operation of V/Stol Aircraft," *Annals of Reliability and Maintainability Conference*, 1966.

15. Drury, C. G., S. G. Schiro, and S. J. Czaja, "Human Reliability in Emergency Medical Response," *Proceedings of the Annual Reliability and Maintainability Symposium*, 1977.

16. Drury, C. G. and J. G. Fox, *Human Reliability in Quality Control*, Halstead, New York, 1976.

17. Fishburn, P., A. Murphy, and H. Isaacs, "Sensitivity of Decisions to Probability Estimation Errors: A Re-Examination," *Oper. Res.*, **16**, 254–267 (1968).

18. Gael, S., "Improving Output Through Job Performance Evaluation," *Annual Reliability and Maintainability Symposium*, 1978.

19. Hagen, E. W., Editor, "Human Reliability Analysis," *Nucl. Safety*, **17**, 315–326 (1976).

20. Halpin, S. M. and E. M. Johnson, "Cognitive Reliability in Manned Systems," *IEEE Trans. Reliability*, **22**, 165–169 (1973).

21. Huston, R. L., "Human Reliability on Man-Machine Interactions," *Annual Reliability and Maintainability Symposium*, 1974.

22. Inaba, K. and R. Matson, "Measurement of Human Errors with Existing Data," *Seventh Annual Reliability and Maintainability Conference*, ASME, New York, 1968.

23. Irwin, I. A., J. J. Levitz, and A. M. Freed, "Human Reliability in the Performance of Maintenance," *Proceedings of the Symposium on Quantification of Human Performance*, Electronic Industries Association and University of New Mexico, Albuquerque, New Mexico, August 1964.

24. Johnson, E. M., R. C. Cavanagh, R. L. Spooner, and M. G. Samet, "Utilization of Reliability Measurements in Bayesian Inference: Models and Human Performance," *IEEE Trans. Reliab.* **R-22**, 1973.

25. Jones, D. M., "The Need for Quantification in Human Factors Engineering," *Sixth Reliability and Maintainability Conference*, 1967.

26. Juran, J. M., "Operator Errors—Time for a New Look," *Qual. Control*, **1**, (1968), pp. 9–11.

27. Katter, R. V., "On Managing the Present Through Efficient Use of the Past," *Sixth Reliability and Maintainability Conference*, 1967.

28. Kaufman, R. A., T. F. Oehrlein, and M. L. Kaufmann, "Predicting Human Reliability— Implications for Operations and Maintenance in Space," National IAS-ARS Joint Meeting, June 13–16, 1961.

29. Keenan, J. J., "Interactionist Models of the Varieties of Human Performance in Complex Work Systems," *Sixth Reliability and Maintainability Conference*, 1967.

30. Kelly, C. W. and S. Barclay, "Improvement of Human Reliability Using Bayesian Hierarchical Inference," *Annual Reliability and Maintainability Symposium*, 1974.

31. Koppa, R. J. and G. G. Hayes, "Determination of Motor Vehicle Characteristics Affecting Driver Handling Performance," *Proceedings of the Annual Reliability and Maintainability Symposium*, 1976.

32. Kraft, J. A., "Mitigating of Human Error Through Human Factors Design Engineering," *Annual Reliability and Maintainability Conference*, ASME, New York, 1968.

33. Kragt, H., "Human Reliability Engineering," *IEEE Trans. Reliability*, **R-27**, 195–201, (1978).

34. Lamb, J. C. and K. E. Williams, "Prediction of Operator Performance for Sonar Maintenance," *IEEE Trans. Reliab.*, **R-22**, 131–134 (1973).

35. Lamb, J. C., "A Test of a Basic Assumption of Human Performance Modeling," *Annual Reliability and Maintainability Symposium*, 1972.

36. Larsen, O. A. and S. I. Sander, "Development of Unit Performance Effectiveness Measures Using DELPHI Procedures," NPRDC-TR-76-12, Navy Research and Development Center, San Diego, CA, September 1975.

37. LaSala, K. P., A. I. Siegel, and C. Sontz, "Allocation of Man-Machine Reliability," *Proceedings of the Annual Reliability and Maintainability Symposium*, 1976.

38. LaSala, K. P., A. I. Siegel, and C. Sontz, "Man-Machine Reliability—A Practical Engineering Tool," *Annual Reliability and Maintainability Symposium*, 1978.

39. Lees, F. P., "Quantification of Man-Machine System Reliability in Process Control," *IEEE Trans. Reliab.*, **R-22**, (1973).

40. Lees, F. P., "Man-Machine System Reliability, In: E. Edwards and F. P. Lees," *Man and Computer in Process Control*, The Institution of Chemical Engineers, London, England, 1973.

41. Lincoln, R. S., "Human Factors in Attainment of Reliability," *IRE Trans. Reliability Qual. Contr.* (1962). pp. 97–103.

42. Lincoln, R. S., "Human Factors in the Attainment of Reliability," *IRE Trans. Reliability Contr.* (1960). pp. 97–103.

43. Majesty, M. S., "Personnel Sub-System Reliability for Aerospace Systems," *Proceedings IAS National Aerospace Systems Reliability Symposium*, 1962.

44. Manz, G. W., "Human Engineering: Aids to Smooth Operation," *Nucl. Safety*, **18**, 223–227 (1977).

45. Meister, D., "A Critical Review of Human Performance Reliability Predictive Methods," *IEEE Trans. Reliab.*, **22**, 116–123 (1973).

46. Meister, D., "Comparative Analysis of Human Reliability Models," AD 734 432, NTIS, Springfield, Virginia, USA 1971.

47. Meister, D., "Subjective Data in Human Reliability Estimates," *Annual Reliability and Maintainability Symposium*, 1978.

48. Meister, D. and R. G. Mills, "Development of Human Performance Reliability Data System," *Annals of Reliability and Maintainability Symposium*, 1972.

49. Meister, D., "Human Factors in Reliability," In: Edited by W. G. Ireson, *Reliability Handbook*, McGraw-Hill, New York, 1966.

50. Meister, D., "The Problem of Human-Initiated Failures," *Eighth National Symposium on Reliability and Quality Control*, 1962.

51. Muller, P. F., "Potential Damage Evaluation: A Method for Determining the Potential for Human-Caused Damage in Operating Systems," *Proceedings of the Reliability and Maintainability Conference*, 1968.

52. Munger, S. J., et al., R. W. Smith, D. Payne, "An Index of Electronic Equipment Operability: Data Store," Report AIR-C43-1/62-RP (1), American Institute for Research, Pittsburgh, PA, January 1962.

53. Nahvi, M. J., "Reliability of Human Visual Signal Detection in the Presence of Noise," *IEEE Trans. Reliability*, **R-23**, 326–331 (1974).

54. Nawrocki, L. H., M. H. Strub, and R. M. Cecil, "Error Categorization and Analysis in Man-Computer Communication Systems," *IEEE Trans. Reliability*, **R-23**, (1978).

55. Page, H. J., "The Human Element in the Maintenance Package," *Eighth National Symposium on Reliability and Quality Control*, 1962.

56. Peters, G. A. and T. A. Hussman, "Human Factors in Systems Reliability," *Human Factors*, **1**, 38–50 (1959).

57. Pontecorvo, A. B., "A Method of Predicting Human Reliability," *Annual Reliability and Maintainability Symposium*, 1965.

58. Rabideau, G. F., "Prediction of Personnel Sub-System Reliability Early in the System Development Cycle," *Proceedings IAS National Aerospace Systems Reliability Symposium*, 1962.

59. Ramsey, J. D., "Reliability and Comparability of Heat Exposure Indices," *Proceedings of the Annual Reliability and Maintainability Symposium*, 1976.

60. Rasmussen, J., "The Role of the Man-Machine Interface in Systems Reliability," *NATO Generic Conference*, Liverpool, England, 1973.

61. Reactor Safety Report, WASH, 1400, Apx III and IV, NTIS, Springfield, IL, 1975.

62. Regulinski, T. L., "Human Performance Reliability Modeling in Time Continuous Domain," *Proceedings NATO Generic Conference*, Liverpool, England, 1973.

63. Regulinski, T. L. and W. B. Askren, "Mathematical Modeling of Human Performance Reliability," *Proceedings of Annual Symposium on Reliability*, 1969.

64. Regulinski, T. L., "Stochastic Modeling of Human Performance Effectiveness Functions," *Annual Reliability and Maintainability Symposium*, 1972.

65. Regulinski, T. L., "On Modeling Human Performance Reliability," *IEEE Trans. Reliab.* **R-22**, 114–115 (1973).

66. Rigby, L. V. and A. D. Swain, "Effects of Assembly Error on Product Acceptability and Reliability," *Proceedings of the Seventh Annual Reliability and Maintainability Conference*, 1968.

67. Rigby, L. V., "Why Do People Drop Things," *Quality Progr.*, 16–19 (1973).

68. Robinson, J. E., W. E. Deutch, and J. G. Rogers, *Human Factors*, **12**, 256–267 (1970).

69. Schum, D. and W. Ducharme, "Comments on the Relationship Between the Impact and the Reliability of Evidence." *Organiz. Behav. Human Perf.*, **6**, (1971).

70. Schum, D. A. and P. E. Pfeiffer, "Observer Reliability and Human Inference," *IEEE Trans. Reliability*, **22**, (1973).

71. Siegel, A. E., J. Jay Wolf, and M. R. Lautman, "A Family of Models for Measuring Human Reliability," *Proceedings of the Annual Reliability and Maintainability Symposium*, 1975.

72. Siegel, A. I., "A Method for Predicting the Probability of Effective Equipment Maintenance," *Annual Reliability and Maintainability Symposium*, 1972.

73. Smith, C. O., *Introduction to Reliability in Design*, McGraw-Hill, New York, 1976.

74. Sontz, C. and J. C. Lamb, "Predicting System Reliability from Human Data," *Proceedings of the Annual Reliability and Maintainability Symposium*, 1975.

75. Sriyananda, H. and D. R. Towill, "Prediction of Human Operator Performance," *IEEE Trans. Reliab.*, **R-22**, 145–156 (1973).

76. Street, R. L., "Reducing Maintenance Error by Human Engineering Techniques," *Proceedings of the Annual Reliability and Maintainability Symposium*, 1974.

77. Swain, A. D. and H. E. Guttmann, "Human Reliability Analysis Applied to Nuclear Power," *Proceedings of the Annual Reliability and Maintainability Symposium*, 1975.

78. Swain, A. D., "Development of a Human Error Rate Data Bank," *Proceedings US Navy Human Reliability Workshop*, NAVSHIPS 0967-412-4010, February 1977.

79. Swain, A. D., "Shortcuts in Human Reliability Analysis, Generic Techniques in Systems Reliability Assessment," Noordhoff, Leyden, 1974.

80. Swain, A. D., "The Human Element in System Development," *Annual Symposium on Reliability*, 1970.

81. Swain, A. D., "Human Factors in Design of Reliable Systems," Sandia Corporation, Report SC-R-748, February 1964.

82. Swain, A. D., "Reliable Systems vs. Automatic Testing," *Proceedings of the Ninth National Symposium on Reliability and Quality Control*, 1963.

83. Swain, A. D., "Overview and Status of Human Factors Reliability Analysis," *Proceedings of the Eighth Reliability and Maintainability Conference*, New York, July 1969.

84. Swain, A. D., "A Method for Performing Human Factor Reliability Analysis," Sandia Corporation, SCR-685, August 1963.

85. Swain, A. D., "Some Problems in the Measurement of Human Performance in Man-Machine Systems," *Human Factors*, **6**, pp. 687–700, (1964).

86. Swain, A. D., "Design of Industrial Jobs a Worker Can and Will Do," *Human Factors*, **15**, 129–136 (1973).

87. Swain, A. D., "An Error-Cause Removal Program for Industry," *Human Factors*, **15**, 207–221 (1973).

88. Teichner, W. H., "Prediction of Human Performance," *Annual Reliability and Maintainability Symposium*, 1972.

89. Thompson, C. W. N., "Model of Human Performance Reliability in Health Care Systems," *Annual Reliability and Maintainability Symposium*, 1974.

90. Topmiller, D. A. and N. M. Aume, "Computer-Graphic Design for Human Performance," *Annual Reliability and Maintainability Symposium*, 1978.

91. Topmiller, D. A., "Human Factors and Systems Effectiveness," *Annals of Reliability and Maintainability Conference*, 1966.

92. Towill, D. R., "Recent Developments in the Prediction of Human Operator Performance," *1973 NATO Generic Studies Conference*, Noordhoff, Leyden, 1976.

93. Urmston, R., "Operational Performance Recording and Evaluation Data System (OPREDS)," Descriptive Brochures, Code 3400, NAVY Electronics Laboratory Center, San Diego, CA, November 1971.

94. Williams, H. L., "Reliability Evaluation of the Human Component in Man-Machine Systems," Electrical Manufacturing, April 1958.

95. Yun, K. W. and F. E. Kalivoda, "A Model for An Estimation of the Product Warranty Return Rate," *Proceedings of the Annual Reliability and Maintainability Symposium*, 1977.

8

Three-State Device Systems

8.1 INTRODUCTION

A three-state device operates satisfactorily in its normal mode but can fail in either of the two other modes. Typical examples of such a device are a fluid flow valve and an electronic diode. Closed (shorted) and open failure modes pertain to such devices.

Redundancy can generally be used to increase the reliability of a system without any change in the reliability of the individual devices that form the system. However, in the case of a system containing three-state devices, redundancy may either increase or decrease the system reliability. This depends upon the dominant mode of component failure, configuration of the system and the number of redundant components.

An electronic diode and a fluid flow valve are typical examples of three-state devices. Either of these components may fail catastrophically in either the open or closed (shorted) mode. A given three-state device will then have a probability of failure in the open-mode and a probability of failure in the closed or shorted mode. Because a three-state device cannot fail simultaneously in both the open and closed (shorted) modes, the failures are mutually exclusive events. The failure of any one such device is considered independent of all the others.

Three-state devices can be arranged in various redundant configurations such as series, parallel, series-parallel, parallel-series, and mixed arrangements. As these configurations become more complex, the analysis of networks becomes more cumbersome, and redundancy can result in decreased overall system reliability. This lower system reliability is due to the redundancy of the dominant adverse mode of failure.

8.2 LITERATURE REVIEW

Careful consideration of the reliability of three-state devices was presented by Moore and Shannon [27] and Creveling [7] in their 1956 papers on electrical and electronic devices. Creveling developed the reliability and failure equations for a diode quad arrangement, whereas Moore and Shannon developed formulas for several relay networks.

The year 1957 brought another development when Lipp [25] discussed the topology of switching elements versus reliability. The following year, Price [29] specifically dealt with the reliability of three-state devices in a parallel configuration and attempted to optimize the number of redundant components. In 1960, Barlow and Hunter [1–3] used calculus to optimize the reliability of series, parallel, series-parallel, and parallel-series networks. They also computed the number of components that maximize the expected system life for these first two types of systems assuming component life is exponentially distributed.

In 1962 Sorensen [35] applied the theory established by the previous researchers on three-state device networks to several electronic circuits. His primary approach was very similar to that of Creveling. In the same year, Cluley [6] published a paper on low-level redundancy as a means of improving the reliability of digital computers. Also in 1962 James et al. [23] reviewed the reliability problem and derived some systems reliability equations for redundant three-state device structures. In 1963, Blake [4] extended the work of Moore and Shannon [27] on networks of relay contacts by investigating the open and short circuit failures of hammock networks. Barlow et al. [3] extended their previous contribution to maximize the expected system life for components having exponential and uniform time to failure distributions.

In 1967 Kolesar [24] extended the work of the previous researchers when he optimized a series-parallel three-state device structure under constrained conditions. In 1970 Misra and Rao [26] developed a signal flow graph approach. During the following 2 years, only one of the four studies making reference to the subject appears to be important. Evans [19] gave a very brief introduction to three-state device reliabilities in his paper and Butler [5] made brief reference to it in his publication.

Since 1975 several contributions on the subject have been made by Dhillon [8–17, 30–34].

8.3 RELIABILITY ANALYSIS OF THREE-STATE DEVICE NETWORKS

The system reliability equations are developed for several configurations in this section. More detailed derivations are described in Appendix.

8.3.1 Series Structure

In a series configuration any one component failing in an open mode causes system failure, whereas all elements of the system must malfunction in a shorted mode for the system to fail. The system reliability is given by (8.1).

$$R_s = \prod_{i=1}^{n} (1 - q_{oi}) - \prod_{i=1}^{n} q_{si} \tag{8.1}$$

where R_s = the series system reliability

   n = the number of nonidentical independent three-state compo-
     nents

   q_{oi} = the probability of open-mode failure of component i

   q_{si} = the probability of short-mode failure of component i

In the case of component constant open and short mode failure rates, the open and short mode failure probability equations become [8]

$$q_o(t) = \frac{\lambda_o}{\lambda_o + \lambda_s} \{ 1 - e^{-(\lambda_o + \lambda_s)t} \} \tag{8.2}$$

and

$$q_s(t) = \frac{\lambda_s}{\lambda_o + \lambda_s} \{ 1 - e^{-(\lambda_o + \lambda_s)t} \} \tag{8.3}$$

where λ_o = the open-mode constant failure rate

   λ_s = the short-mode constant failure rate

   t = time

The derivation of (8.2) and (8.3) are shown in Section 8.5.2. To obtain (8.2) and (8.3) set $\mu_1 = \mu_2 = 0$ in (8.57) and (8.58), respectively. By substituting expressions (8.2) and (8.3) in (8.1) we get:

$$R(t) = \prod_{i=1}^{n} \left[1 - \frac{\lambda_{oi}}{\lambda_{oi} + \lambda_{si}} \{ 1 - e^{-(\lambda_{oi} + \lambda_{si})t} \} \right]$$

$$- \prod_{i=1}^{n} \frac{\lambda_{si}}{\lambda_{oi} + \lambda_{si}} \{ 1 - e^{-(\lambda_{oi} + \lambda_{si})t} \} \tag{8.4}$$

Short Failure Mode Probability.   The system short or closed failure mode probability, Q_s, is given by

$$Q_s = \prod_{i=1}^{n} q_{si} \tag{8.5}$$

Open Failure Mode Probability.   Probability of open mode failure for a series system is given by

$$Q_o = 1 - \prod_{i=1}^{n} (1 - q_{oi}) \tag{8.6}$$

Where Q_o is the probability of open mode failure of series network.

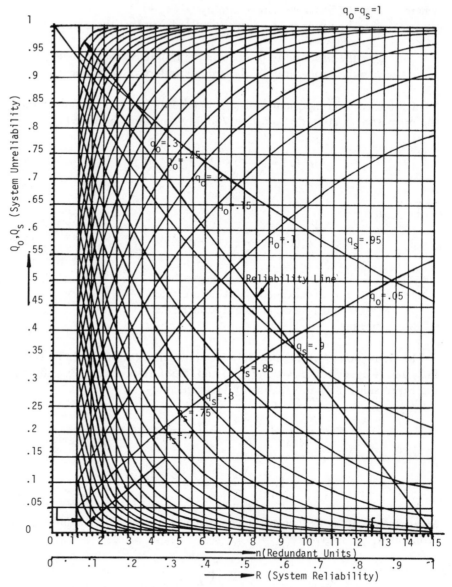

Figure 8.1 An identical component series structure unreliability plot.

Plots of (8.5) and (8.6) are shown in Figure 8.1. This figure shows that the open mode failure probability increases as the number of redundant components in the series system increases.

Example 1. Consider two independent identical diodes connected in series. Open and short circuit failure probabilities are 0.2 and 0.1, respectively. It is required to find the system reliability of the two diodes for this simple arrangement.

In this case $n=2$, $q_s=0.1$ and $q_o=0.2$. Rewrite (8.1) for two identical diodes

$$R_s = (1-q_o)^2 - q_s^2 \tag{8.7}$$

For given data,

$$R_s = (1-0.2)^2 - (0.1)^2 = 0.63$$

8.3.2 Parallel Structure

For a parallel configuration, all the elements must fail in the open-mode or any one of the elements must be in a short-mode to cause the system to fail. The parallel network reliability is given by

$$R = \prod_{i=1}^{m} (1-q_{si}) - \prod_{i=1}^{m} q_{oi} \tag{8.8}$$

where m is the number of nonidentical independent elements.

The open and short failure mode probability plots are the same as shown in Figure 8.1. Because of duality, the short failure mode probability replaces the open failure probability and vice versa. The same duality concept applies to (8.1) and (8.8).

Example 2. Suppose the data of Example 1 is used for parallel configuration; evaluate the system reliability by using (8.8)

$$R = (1-q_s)^2 - q_0^2 = (1-0.1)^2 - (0.2)^2 = 0.77$$

The parallel system reliability is 0.77.

8.3.3 Series-Parallel Network

This is a combination of series and parallel configurations. System reliability is given by (8.9) for n identical independent units, each containing m independent elements:

$$R = \left\{ 1 - \prod_{i=1}^{m} q_{oi} \right\}^n - \left\{ 1 - \prod_{i=1}^{m} (1-q_{si}) \right\}^n \tag{8.9}$$

Example 3. Consider the reliability evaluation of series-parallel arrays of the identical fluid flow valves with $q_o=0.2$, $q_s=0.1$, $n=2$ and $m=4$.

For $n=2$ and $m=4$ (8.9) becomes

$$R=\left(1-q_o^4\right)^2-\left\{1-(1-q_s)^4\right\}^2 \tag{8.10}$$

For $q_s=0.1$, $q_o=0.2$, the system reliability

$$R=(1-0.2^4)^2-\left\{1-(1-0.1)^4\right\}^2=0.88$$

8.3.4 Parallel-Series Structure

This configuration is a dual of the series-parallel network. The system reliability equation for a configuration containing m identical units and n number of nonidentical series elements becomes

$$R=\left(1-\prod_{i=1}^{n}q_{si}\right)^m-\left[1-\prod_{i=1}^{n}(1-q_{oi})\right]^m \tag{8.11}$$

Example 4. Use the date given in Example 3 and evaluate the parallel-series network reliability. Therefore

$$R=\left(1-q_s^2\right)^4-\left\{1-(1-q_o)^2\right\}^4$$
$$=(1-0.1^2)^4-\left\{1-(1-0.2)^2\right\}^4 \tag{8.12}$$
$$=0.9438$$

8.3.5 Bridge Network

This configuration is shown in Figure 8.2. The following bridge reliability equation, R_b is taken from reference 25:

$$R_b=1-Q_{o1}-Q_{o2} \tag{8.13}$$

where Q_{ok} is the network open failure mode probability, for $k=1$
        Q_{ok} is the network short (close) failure mode probability, for $k=2$

and

$$Q_{OK}=2\prod_{i=1}^{5}\Phi_i-\prod_{i=2}^{5}\Phi_i-\prod_{\substack{i=1\\i\neq2}}^{5}\Phi_i-\prod_{\substack{i=1\\i\neq3}}^{5}\Phi_i-\prod_{\substack{i=1\\i\neq4}}^{5}\Phi_i-\prod_{i=1}^{4}\Phi_i$$

$$+\prod_{\substack{i=1\\i\neq2,4}}^{5}\Phi_i+\prod_{i=2}^{4}\Phi_i+\prod_{\substack{i=1\\i\neq2,3}}^{4}\Phi_i+\prod_{\substack{i=2\\i\neq3,4}}^{5}\Phi_i \tag{8.14}$$

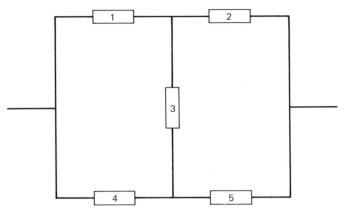

Figure 8.2 A bridge network of dissimilar components.

for

$$K = \begin{cases} 1, & \Phi_i = q_{oi} \\ 2, & \Phi_i = q_{si} \end{cases}$$

As shown in Figure 8.2, the bridge network is composed of five elements, $i = 1, 2, \ldots, 5$, where the element number 3 is known as the critical element.

8.4 DELTA-STAR TRANSFORMATION TECHNIQUE

The reliability evaluation of series, parallel, and series-parallel networks is widely discussed. To evaluate the reliability of a bridge, or other such complex structures, the theories in the literature are difficult to apply. The delta-star transformation [8] is a simple approach for such problems. This technique transforms a complex structure to a series and parallel form. Thereon the network reduction technique may be applied to obtain reliability of transformed configuration. The technique introduces a small error, which can be neglected for practical purposes.

Transformations are carried out in terms of both of the failure modes instead of simply reliability or unreliability as is the case for a two-state device structure.

The resulting delta-star transformation formulas are developed by finding the leg equivalent, as illustrated by Figure 8.3.

8.4.1 Open-Failure Mode

The delta-star leg equivalents are obtained in the same manner as the simpler two-state component case. Figure 8.4 illustrates the leg equivalents for the open-mode failure case.

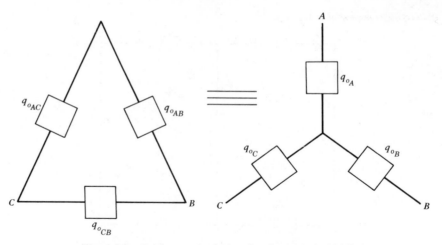

Figure 8.3 A delta-star equivalent for the open-mode failure.

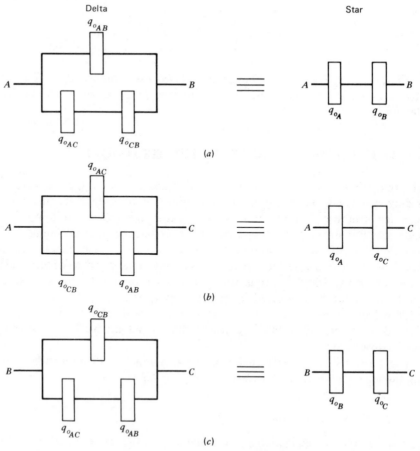

Figure 8.4 Delta-star equivalent legs.

Again, by using the independent probability laws for the series and parallel structures, the equivalent legs of the block diagrams as shown in Figure 8.4a, b, and c result in (8.17), (8.18), and (8.19), respectively:

Series Case. Series system open-mode unreliability

$$Q_o = 1 - \prod_{i=1}^{n} (1 - q_{oi}) \tag{8.15}$$

Where q_{oi} is the components' open-mode unreliability, $i = 1, n$.

Parallel Structure Case. Open-mode system unreliability

$$Q_o = \prod_{i=1}^{n} q_{oi} \tag{8.16}$$

With the aid of (8.15) and (8.16) the equivalent legs of the block diagrams are transformed, respectively, to the following:

$$1 - (1 - q_{o_A})(1 - q_{o_C}) = \left[1 - (1 - q_{o_{CB}})(1 - q_{o_{AB}}) \right] q_{o_{AC}} \tag{8.17}$$

$$1 - (1 - q_{o_A})(1 - q_{o_B}) = \left[1 - (1 - q_{o_{AC}})(1 - q_{o_{CB}}) \right] q_{o_{AB}} \tag{8.18}$$

$$1 - (1 - q_{o_B})(1 - q_{o_C}) = \left[1 - (1 - q_{o_{AC}})(1 - q_{o_{AB}}) \right] q_{o_{CB}} \tag{8.19}$$

From these simultaneous equations result the following delta-star conversion equations:

$$q_{o_A} = 1 - \left[\frac{\left[1 - \{ 1 - (1 - q_{o_{CB}})(1 - q_{o_{AB}}) \} q_{o_{AC}} \right]\left[1 - \{ 1 - (1 - q_{o_{AC}})(1 - q_{o_{CB}}) \} q_{o_{AB}} \right]}{\left[1 - \{ 1 - (1 - q_{o_{AC}})(1 - q_{o_{AB}}) \} q_{o_{CB}} \right]} \right]^{1/2} \tag{8.20}$$

$$q_{o_B} = 1 - \left[\frac{\left[1 - \{ 1 - (1 - q_{o_{AC}})(1 - q_{o_{CB}}) \} \right]\left[1 - \{ 1 - (1 - q_{o_{AC}})(1 - q_{o_{AB}}) \} q_{o_{CB}} \right]}{\left[1 - \{ 1 - (1 - q_{o_{CB}})(1 - q_{o_{AB}}) \} q_{o_{AC}} \right]} \right]^{1/2} \tag{8.21}$$

$$q_{o_C} = 1 - \left[\frac{\left[1 - \{ 1 - (1 - q_{o_{CB}})(1 - q_{o_{AB}}) \} q_{o_{AC}} \right]\left[1 - \{ 1 - (1 - q_{o_{AC}})(1 - q_{o_{AB}}) \} q_{o_{CB}} \right]}{\left[1 - \{ 1 - (1 - q_{o_{AC}})(1 - q_{o_{CB}}) \} q_{o_{AB}} \right]} \right]^{1/2} \tag{8.22}$$

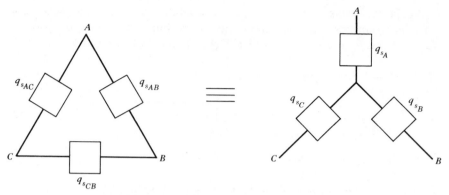

Figure 8.5 A short-failure delta-star transformation.

8.4.2 Short-Failure Mode

Similarly, as for the open-failure mode, Figures 8.5 and 8.6 show the short-failure mode equivalent configurations.

Again, with aid of the independent probability laws for parallel and series structures, (8.25)–(8.27) are obtained from their equivalent corresponding legs of the block diagrams of Figure 8.6a–c.

Series Case.   System short-mode unreliability

$$Q_s = \prod_{i=1}^{n} q_{si} \tag{8.23}$$

where q_{si} is the components' short-mode unreliability, $i = 1, n$.

Parallel Structure Case.

$$Q_s = 1 - \prod_{i=1}^{n} (1 - q_{si}) \tag{8.24}$$

With applications of (8.23) and (8.24) to the equivalent legs of the block diagrams of Figure 8.6a–c the corresponding equations become

$$q_{s_A} q_{s_C} = 1 - (1 - q_{s_{CB}} q_{s_{AB}})(1 - q_{s_{AC}}) \tag{8.25}$$

$$q_{s_A} q_{s_B} = 1 - (1 - q_{s_{AC}} q_{s_{CB}})(1 - q_{s_{AB}}) \tag{8.26}$$

$$q_{s_B} q_{s_C} = 1 - (1 - q_{s_{AC}} q_{s_{AB}})(1 - q_{s_{CB}}) \tag{8.27}$$

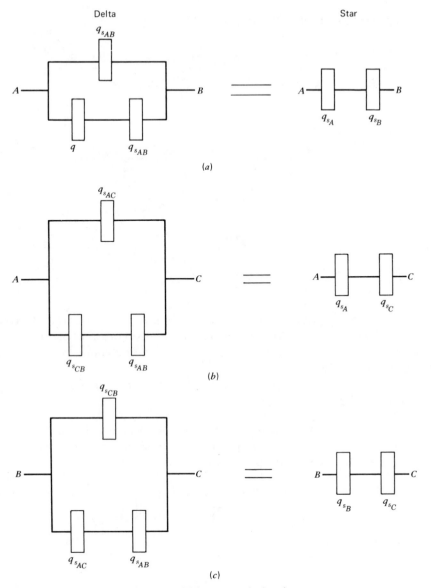

Figure 8.6 Delta-star equivalent legs.

Solving these equations simultaneously yields

$$q_{s_A} = \left[\frac{\left[1-(1-q_{s_{CB}}q_{s_{AB}})(1-q_{s_{AC}})\right]\left[1-(1-q_{s_{AC}}q_{s_{CB}})(1-q_{s_{AB}})\right]}{\left[1-(1-q_{s_{AC}}q_{s_{AB}})(1-q_{s_{CB}})\right]} \right]^{1/2}$$

$$(8.28)$$

$$q_{s_B} = \left[\frac{\left[1-(1-q_{s_{AC}}q_{s_{AB}})(1-q_{s_{CB}})\right]\left[1-(1-q_{s_{AC}}q_{s_{CB}})(1-q_{s_{AB}})\right]}{\left[1-(1-q_{s_{CB}}q_{s_{AB}})(1-q_{s_{AC}})\right]} \right]^{1/2}$$

$$(8.29)$$

$$q_{s_C} = \left[\frac{\left[1-(1-q_{s_{CB}}q_{s_{AB}})(1-q_{s_{AC}})\right]\left[1-(1-q_{s_{AC}}q_{s_{AB}})(1-q_{s_{CB}})\right]}{\left[1-(1-q_{s_{AC}}q_{s_{CB}})(1-q_{s_{AB}})\right]} \right]^{1/2}$$

$$(8.30)$$

It is readily seen that (8.28), (8.29), and (8.30) are all interrelated. After computing the unreliability value by use of the first equation, the computation for the other two is made easier because the first computation is used in their evaluation.

The same sort of argument applies to the open-failure equations (8.20), (8.21), and (8.22).

Example 5. A bridge network example is solved here to illustrate the use of these formulas. As an example, the network shown by Figure 8.7 is one where the delta configuration is identified with the labels *A*, *B*, and *C*.

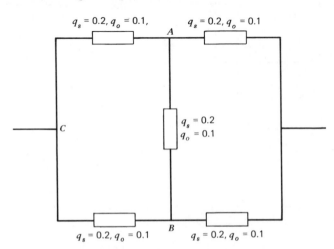

Figure 8.7 A three-state bridge structure.

Its equivalent open and short failure mode probability values for this situation are obtained by using (8.20)–(8.22) and (8.28)–(8.30), respectively. The numerical results obtained are as follows:

Open-mode failure probability:

$$q_{o_A} = 0.01 \qquad q_{o_B} = 0.01 \qquad q_{o_C} = 0.01$$

and

$$q_{s_A} = 0.482 \qquad q_{s_B} = 0.482 \qquad q_{s_C} = 0.482$$

These relationships allow Figure 8.7 to be redrawn as its equivalent as shown by Figure 8.8. The resulting total open and short mode probabilities of failure for Figure 8.8 are

$$Q_o = 1 - \left[1 - \left\{ (1 - q_{s_1})(1 - q_{o_A}) \right\} \left\{ 1 - (1 - q_{o_B})(1 - q_{o_2}) \right\} \right] \left[1 - q_{o_C} \right]$$

(8.31)

and

$$Q_s = \left[1 - (1 - q_{s_1} q_{s_A})(1 - q_{s_2} q_{s_B}) \right] q_{s_C}$$

(8.32)

By using (8.31) and (8.32)

$$Q_o = 0.022 \qquad Q_s = 0.088$$

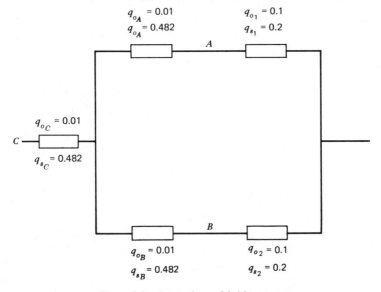

Figure 8.8 A transformed bridge structure.

thereby giving bridge reliability

$$R_T = 1 - Q_o - Q_s = 0.89$$

8.5   REPAIRABLE THREE-STATE DEVICE SYSTEMS

This section presents several mathematical models of repairable systems. Most of these models are available in the referenced literature.

8.5.1   *Analysis of a Three-State System with Two Types of Components*

This model is developed by using the supplementary variables technique [39–41]. The three-state model [9] discussed in this section is shown in Figure 8.9. The components of this system are divided into classes (i.e., Class I and II). If any one component of Class I fails, the system will experience a complete system failure. A component failure of Class II will cause a catastrophic system failure. Some typical examples of such a system are automatic machines, fluid flow valves, a rotational mechanical system that jams so that rotation is blocked, a shaft that shears so that an input rotation causes no output rotation, and an electrical or electronic system.

   System states are defined as follows:

1. *Normal state*. The successful functioning of a device.
2. *Complete failure state*. Total system failure (i.e., the machine does not operate at all), normally caused by the failure of a Class I component.
3. *Catastrophic failure state*. The system failure state in which a system or equipment carries out unacceptable operations, usually caused by the failure of a Class II component.

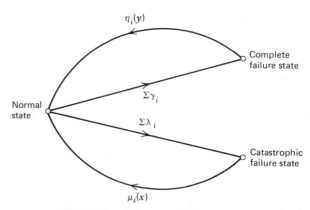

Figure 8.9   A three-state Markov model.

Suppose an automatic machine carries out some operations on assembly line items. The automatic machine is composed of many component parts; therefore, the components of the machine can be divided into two classes (i.e., Class I and II). A component failure of Class I causes the complete failure or breakdown of the automatic machine. A failure of any one component of Class II will cause a catastrophic failure of the automatic system (this type of failure will initiate some unwanted operations on the assembly line items).

Obviously, to restore the automatic machine back to its normal state, repair is necessary. Repair times are arbitrarily distributed.

The following notations and definitions are used to formulate this Markov Model:

$P_o(t)$ = the probability of the system being in its normal mode at time t.

$P_{1,i}(y, t)\Delta$ = the probability that at time t, the system which has failed, because of the failure of its ith component in Class I, is being repaired and the elapsed repair time lies in the interval $(y, y + \Delta t)$ for $i = 1, 2, 3, n$.

$P_{2,i}(x, t)\Delta$ = the probability that at time t the system that has failed, because of the failure of its ith component in Class II, is being repaired and the elapsed repair time lies in the interval $(x, x + \Delta t)$ for $1 = 1, 2, 3, n$.

$\eta_i(y)\Delta \equiv$ the first-order probability, that the ith component of Class I is repaired in the interval $(y, y + \Delta)$, conditioned that it was not repaired up to time y.

$\mu_i(x)\Delta \equiv$ the first-order probability, that the ith component of Class II is repaired in the interval $(x, x + \Delta)$, conditioned that it was not repaired up to time x.

λ_i = the constant failure rate of the ith component of Class II.

γ_i = the constant failure rate of the ith component of Class I.

s = the Laplace transform variable.

Assumptions

1. Failures are statistically independent.
2. A failed system is restored as good as new.

A Mathematical Model. The integro-differential equations (and associated boundary-initial conditions) associated with Figure 8.9 are

$$\left[\frac{\partial}{\partial t} + \lambda + \gamma \right] \cdot P_0(t) = \sum_{i=1}^{n} \int_0^\infty P_{2,i}(x, t)\mu_i(s)\, dx$$

$$+ \sum_{i=1}^{n} \int_0^\infty P_{1,i}(y, t)\eta_i(y)\, dy \qquad (8.33)$$

$$\left[\frac{\partial}{\partial t}+\frac{\partial}{\partial y}+\eta_i(y)\right]\cdot P_{1,i}(y,t)=0 \tag{8.34}$$

$$\left[\frac{\partial}{\partial t}+\frac{\partial}{\partial x}+\mu_i(x)\right]\cdot P_{2,i}(x,t)=0 \tag{8.35}$$

$$P_{1,i}(0,t)=\gamma_i\cdot P_0(t) \qquad P_{2,i}(0,t)=\lambda_i\cdot P_0(t)$$

$P_0(0)=1$, at $t=0$ other initial condition probabilities are zero, where

$$\lambda=\sum_{i=1}^{n}\lambda_i \qquad \gamma=\sum_{i=1}^{n}\gamma_i$$

Solving the above integro-differential equations by Laplace transforms and integration (including some substitutions) will yield

$$P_0(s)=\frac{1}{\left[(s+\lambda+\gamma)-\sum_{i=1}^{n}\lambda_i G_{2,i}(s)-\sum_{i=1}^{n}\gamma_i G_{1,i}(s)\right]} \tag{8.36}$$

where

$$G_{1,i}(s)=\int_0^{\infty}e^{-sy}\eta_i(y)\exp\left(-\int_0^{y}\eta_i(y)\,dy\right)dy$$

$$G_{2,i}(s)=\int_0^{\infty}e^{-sx}\mu_i(s)\exp\left(-\int_0^{x}\mu_i(x)\,dx\right)dx$$

Since

$$P_{1,i}(s)=\int_0^{\infty}P_{1,i}(y,s)\,dy \tag{8.37}$$

$$P_{2,i}(s)=\int_0^{\infty}P_{2,i}(x,s)\,dx, \tag{8.38}$$

where $P_{1,i}(s)$, $P_{2,i}(s)$ are the Laplace transform of probabilities $P_{1,i}(t)$, $P_{2,i}(t)$ that the system is under repair due to the failure of the ith component in Classes I and II, respectively. Therefore,

$$P_{j,i}(s)=P_0(s)\left\{\frac{1-G_{j,i}(s)}{s}\right\}k_j \tag{8.39}$$

for $j=1,2;\ i=1,2,3,n;\ k_1=\gamma_i;\ k_2=\lambda_i$

The Laplace transforms of probabilities $P_1(t)$ and $P_2(t)$, that system is under repair due to the failure of any one of Classes I and II components, respectively, are

$$P_j(s) = \sum_{i=1}^{n} P_{j,i}(s) \qquad \text{for} \quad j = 1, 2 \qquad (8.40)$$

Substituting (8.39) into (8.40) yields

$$P_j(s) = \sum_{i=1}^{n} P_0(s) \left\{ \frac{1 - G_{j,i}(s)}{s} \right\} kj$$

$$\text{for} \quad j = 1, 2; \ k_1 = \gamma_i; \ k_2 = \lambda_i \qquad (8.41)$$

Therefore, for given repair probability density functions $G_{j,i}(t)$, the state probabilities $P_0(t)$, $P_j(t)$ can be obtained by simply taking the inverse Laplace transform of (8.36) and (8.41), respectively.

The steady-state solution, if it exists, of (8.36) and (8.41) can be obtained by employing Abel's Theorem to Laplace transform,

$$\lim_{s \to 0} sf(s) = \lim_{t \to \infty} f(t). \qquad (8.42)$$

Mean time to system failure (MTSF) (if exists) can be obtained from

$$\text{MTSF} = \lim_{s \to 0} P_0(s) \qquad (8.43)$$

More detailed analysis of similar models using the method of supplementary variables are presented in references 39–41.

8.5.2 A Repairable Three-State Device with Constant Failure and Repair Rates

This model [30] is a special case of the model presented in Section 8.5.1. The system transition diagram is shown in Figure 8.10.

Assumptions

1. Failures are statistically independent.
2. The repaired system is as good as new.
3. Repair and failure rates are constant.

Notation

$P_i(t)$ = the probability of the state in question, at time t; $i = 0, 1, 2$
λ = the constant failure rate in question
μ = the constant repair rate in question

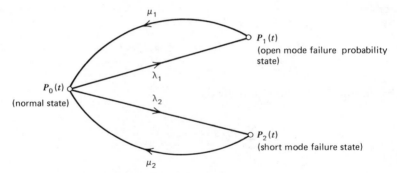

Figure 8.10   A repairable Markov model.

From Figure 8.10, the resulting differential equations are:

$$\frac{dP_0(t)}{dt} = -(\lambda_1 + \lambda_2)P_0(t) = \mu_1 P_1(t) + \mu_2 P_2(t) \tag{8.44}$$

$$\frac{dP_1(t)}{dt} = P_0(t)\lambda_1 - \mu_1 P_1(t) \tag{8.45}$$

$$\frac{dP_2(t)}{dt} = \lambda_2 P_0(t) - \mu_2 P_2(t) \tag{8.46}$$

$$P_0(0) = 1 \qquad P_1(0) = P_2(0) = 0$$

The Laplace transform of (8.44)–(8.46) yields

$$(s + \lambda_1 + \lambda_2)P_0(s) - \mu_1 P_1(s) - \mu_2 P_2(s) = 1 \tag{8.47}$$

$$-\lambda_1 P_0(s) + (s + \mu_1)P_1(s) + 0P_2(s) = 0 \tag{8.48}$$

$$-\lambda_2 P_0(s) + 0P_1(s) + (s + \mu_2)P_2(s) = 0 \tag{8.49}$$

The coefficient of the above simultaneous equations can be written as follows:

$$\left. \begin{array}{ccc} (s + \lambda_1 + \lambda_s) & -\mu_1 & -\mu_2 \\ -\lambda_1 & (s + \mu_1) & 0 \\ -\lambda_2 & 0 & (s + \mu_2) \end{array} \right| \begin{array}{c} 1 \\ 0 \\ 0 \end{array}$$

∴The solution by Cramer's rule yields:

$$P_0(s) = \frac{(s+\mu_1)(s+\mu_2)}{s\left[s^2 + s(\mu_1+\mu_2+\lambda_1+\lambda_2) + (\mu_1\mu_2+\lambda_1\mu_2+\lambda_1\mu_2+\lambda_2\mu_1)\right]}$$

$$(8.50)$$

$$P_1(s) = \frac{\lambda_1(s+\mu_1)}{s\left[s^2 + s(\mu_1+\mu_2+\lambda_1+\lambda_2) + \mu_1\mu_2+\lambda_1\mu_2+\lambda_2\mu_1\right]} \qquad (8.51)$$

$$P_2(s) = \frac{\lambda_2(s+\mu_1)}{s\left[s^2 + s(\mu_1+\mu_2+\lambda_1+\lambda_2) + (\mu_1\mu_2+\lambda_1\mu_2+\lambda_2\mu_1)\right]} \qquad (8.52)$$

The roots of the denominators of (8.50)–(8.52) become

$$k_1, k_2$$

$$= \frac{-(\mu_1+\mu_2+\lambda_1+\lambda_2) \pm \sqrt{(\mu_1+\mu_2+\lambda_1+\lambda_2)^2 - 4(\mu_1\mu_2+\lambda_1\mu_2+\lambda_2\mu_2)}}{2}$$

Now, (8.50)–(8.52) can be expanded in a partial fraction form

$$P_0(s) = \frac{(s+\mu_1)(s+\mu_2)}{s(s-k_1)(s-k_2)}$$

$$= \frac{\mu_1\mu_2}{k_1k_2}\frac{1}{s} + \frac{(k_1+\mu_1)(k_1+\mu_2)}{k_1(k_1-k_2)}\frac{1}{s-k_1} - \frac{(k_2+\mu_1)(k_2+\mu_2)}{k_2(k_1-k_2)}\frac{1}{(s-k_2)}$$

$$(8.53)$$

$$P_1(s) = \frac{\lambda_1(s+\mu_2)}{s(s-k_1)(s-k_2)}$$

$$= \frac{\lambda_1\mu_2}{k_1k_2} + \frac{(\lambda_1k_1+\lambda_1\mu_2)}{k_1(k_1-k_2)}\frac{1}{(s-k_1)} - \frac{(\mu_2+k_2)\lambda_1}{k_2(k_1-k_2)}\frac{1}{(s-k_2)} \qquad (8.54)$$

$$P_2(s) = \frac{\lambda_2(s+\mu_1)}{s(s-k_1)(s-k_2)}$$

$$= \frac{\lambda_2\mu_1}{k_1k_2} + \frac{(\lambda_2k_1+\lambda_2\mu_1)}{k_1(k_1-k_2)}\frac{1}{(s-k_1)} - \frac{(\mu_1+k_2)\lambda_2}{k_2(k_1-k_2)}\frac{1}{(s-k_2)} \qquad (8.55)$$

In time domain, (8.53) and (8.55) become

$$P_0(t) = \frac{\mu_1\mu_2}{k_1k_2} + \left\{ \frac{(k_1+\mu_1)(k_1+\mu_2)}{k_1(k_1-k_2)} \right\} e^{k_1 t} - \left\{ \frac{(k_2+\mu_1)(k_2+\mu_2)}{k_2(k_1-k_2)} \right\} e^{k_2 t}$$

$$(8.56)$$

$$P_1(t) = \frac{\lambda_1\mu_2}{k_1k_2} + \left\{ \frac{\lambda_1 k_1 + \lambda_1\mu_2}{k_1(k_1-k_2)} \right\} e^{k_1 t} - \left\{ \frac{(\mu_2+k_2)\lambda_1}{k_2(k_1-k_2)} \right\} e^{k_2 t} \qquad (8.57)$$

$$P_2(t) = \frac{\lambda_2\mu_1}{k_1k_2} + \left\{ \frac{\lambda_2 k_1 + \lambda_2\mu_1}{k_1(k_1-k_2)} \right\} e^{k_1 t} - \frac{(\mu_1+k_2)\lambda_2}{k_2(k_1-k_2)} e^{k_2 t} \qquad (8.58)$$

Since

$$k_1 k_2 = \mu_1\mu_2 + \lambda_1\mu_2 + \lambda_2\mu_1$$

$$k_1 + k_2 = -(\mu_1 + \mu_2 + \lambda_1 + \lambda_2)$$

therefore, the addition of (8.56)–(8.58) will yield unity, that is,

$$P_0(t) + P_1(t) + P_2(t) = 1$$

The equipment availability is

$$\text{Availability} = P_0(t) = \frac{\mu_1\mu_2}{k_1k_2} + \left\{ \frac{(k_1+\mu_1)(k_1+\mu_2)}{k_1(k_1-k_2)} \right\} e^{k_1 t}$$

$$- \left\{ \frac{(k_2+\mu_1)(k_2+\mu_2)}{k_2(k_1-k_2)} \right\} e^{k_2 t}$$

The availability expression is valid if and only if k_1 and k_2 are negative. As t becomes very large, the steady-state availability equation can be expressed as

$$\lim_{t \to \infty} P_0(t) = \frac{\mu_1\mu_2}{k_1k_2} \qquad (8.59)$$

8.5.3 A Mixed Markov Model with Two Three-State Devices (Master-Slave Relationship)

This mixed Markov model [34] has the two units modeled in series. One device has normal, partial, and catastrophic states and the other has normal, open, and closed mode states (Type II). Repairs are performed only when an equipment fails in its partial mode.

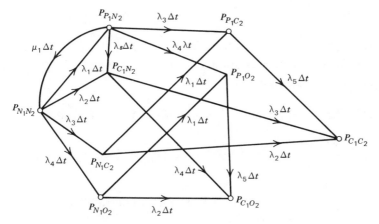

Figure 8.11 A master-slave Markov model.

A typical example of such a system is a fluid flow valve commanded from an instrumentation control panel where the control panel represents the first type of device (Master) and the fluid flow valve represents the second type (Slave). Such practical examples are numerous and may often be encountered in a modern electrical power station. The transition diagram for this case is shown in Figure 8.11.

Abbreviations and Notations

$P(t)$ = probability of the state in question, at time t

$(\cdot)(\cdot)$

N_i = normal mode state of the three-state devices (i.e., master and slave), respectively, $i = 1, 2$.

C_1 = catastrophic failure state of the "master" three-state device

C_2 = closed mode failure state of the "slave" three-state device

P_1 = partial failure state of the "master" three-state device

O_2 = open mode failure state of the "slave" three-state device

λ_1 = constant partial failure rate of the "master" three-state device

λ_2 = constant catastrophic failure rate of the "master" three-state device

λ_s = constant failure rate from partial to catastrophic failure state of the "master" three-state device

λ_3 = constant close mode failure rate of the slave three-state device

λ_4 = constant open mode failure rate of the slave three-state device

μ_1 = constant repair rate of the master device

t = time

Δt = time interval

Assumptions

1. Failures are statistically independent.
2. The repaired system is as good as new.
3. Failure and repair rates are constant.

The state differential equations resulting from Figure 8.11 are

$$\frac{dP_{N_1N_2}(t)}{dt} + (\lambda_1+\lambda_2+\lambda_3+\lambda_4)P_{N_1N_2}(t) = \mu_1 P_{P_1N_2}(t) \qquad (8.60)$$

$$\frac{dP_{P_1N_2}(t)}{dt} + (\lambda_3+\lambda_4+\lambda_5+\mu_1)P_{P_1N_2}(t) = \lambda_1 P_{N_1N_2}(t) \qquad (8.61)$$

$$P_{N_1N_2}(0) = 1 \qquad P_{P_1N_2}(0) = 0$$

Solving the above differential Equations by Laplace transform yields

$$P_{P_1N_2}(t) = \left\{ \frac{\lambda_1}{k_2-k_1} \right\} e^{k_2 t} - \left\{ \frac{\lambda_1}{k_2-k_1} \right\} e^{k_1 t} \qquad (8.62)$$

$$P_{N_1N_2}(t) = \left\{ 1 - \frac{(\lambda_3+\lambda_4+\lambda_5+\mu_1+k_2)}{(k_2-k_1)} \right\} e^{k_1 t}$$

$$+ \frac{\lambda_3+\lambda_4+\lambda_5+\mu_1+k_2}{k_2-k_1} e^{k_2 t} \qquad (8.63)$$

where

$$k_1, k_2 = \frac{-N \pm \sqrt{N^2 - 4AM}}{2A}$$

and

$$A = 1$$

$$N = \lambda_1 + \lambda_2 + 2\lambda_3 + 2\lambda_4 + \lambda_5 + \mu_1$$

$$M = \lambda_1\lambda_3 + \lambda_2\lambda_3 + \lambda_3^2 + 2\lambda_3\lambda_4 + \lambda_1\lambda_4 + \lambda_2\lambda_4 + \lambda_4^2 + \lambda_1\lambda_5 + \lambda_2\lambda_5$$

$$+ \lambda_3\lambda_5 + \lambda_4\lambda_5 + \lambda_2\mu_1 + \lambda_3\mu_1 + \lambda_4\mu_1$$

Therefore,

$$\text{System reliability} = P_{P_1N_2}(t) + P_{N_1N_2}(t) \qquad (8.64)$$

8.5.4 A Repairable Markov Model of Two Units in Series I

Consider two three-state devices arranged in a series configuration [34]. The repair is performed only when one of the devices fails in its closed mode, assuming the other one is still operating. Two fluid flow valves operating in series represent a good example. The transition diagram is shown in Figure 8.12.

Abbreviations and Notations

$P(t)$ = probability of state in question, at time t
$(\cdot)(\cdot)$
$\quad N_i$ = normal mode state of the both three-state devices, $i = 1, 2$
$\quad C_1$ = close mode failure state of the first three-state device
$\quad C_2$ = close mode failure state of the second three-state device
$\quad \mu_1$ = constant repair rate of the first three-state device
$\quad \mu_2$ = constant repair rate of the second three-state device
$\quad \lambda_1$ = constant close mode failure rate of the first three-state device
$\quad \lambda_2$ = constant close mode failure rate of the second three-state
$\quad\quad$ = device
$\quad \lambda_3$ = constant open mode failure rate of the first three-state device
$\quad \lambda_4$ = constant open mode failure rate of the second three-state device
$\quad t$ = time
$\quad \Delta t$ = time interval
$\quad s$ = Laplace Transform variable

Assumptions

1. Failures are statistically independent.
2. Repaired device is good as new.
3. Failure and repair rates are constant.

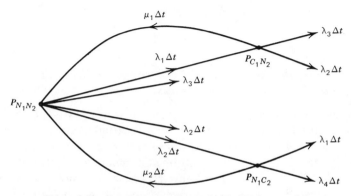

Figure 8.12 A series two-unit repairable Markov model.

The differential Equations associated with Figure 8.12 are

$$\frac{dP_{N_1N_2}(t)}{dt} + (\lambda_1 + \lambda_2 + \lambda_3 + \lambda_4)P_{N_1N_2}(t) = P_{C_1N_2}(t)\mu_1 + P_{N_1C_2}(t)\mu_2 \qquad (8.65)$$

$$\frac{dP_{C_1N_2}(t)}{dt} + (\lambda_2 + \lambda_3 + \mu_1)P_{C_1N_2}(t) = P_{N_1N_2}(t)\lambda_1 \qquad (8.66)$$

$$\frac{dP_{N_1C_2}(t)}{dt} + (\lambda_1 + \lambda_4 + \mu_2)P_{N_1C_2}(t) = P_{N_1N_2}(t)\lambda_2 \qquad (8.67)$$

$$P_{N_1N_2}(0) = 1 \qquad P_{C_1N_2}(0) = 0 \qquad P_{N_1C_2}(0) = 0$$

The values of $P_{N_1N_2}(s)$, $P_{N_1C_2}(s)$, $P_{C_1N_2}(s)$ are obtained from the above differential equations:

$$P_{N_1N_2}(s) = \frac{(s + \lambda_2 + \lambda_3 + \mu_1)(s + \lambda_1 + \lambda_4 + \mu_2)}{\Delta} \qquad (8.68)$$

where $\Delta = \begin{vmatrix} (s + \lambda_1 + \lambda_2 + \lambda_3 + \lambda_4) & -\mu_1 & -\mu_2 \\ -\lambda_1 & (s + \lambda_2 + \lambda_3 + \mu_1) & 0 \\ -\lambda_2 & 0 & (s + \lambda_1 + \lambda_4 + \lambda_2) \end{vmatrix}$

$$P_{C_1N_2}(s) = \frac{\lambda_1(s + \lambda_1 + \lambda_4 + \mu_2)}{\Delta} \qquad (8.69)$$

$$P_{N_1C_2}(s) = \frac{-\lambda_2(s + \lambda_2 + \lambda_3 + \mu_1)}{\Delta} \qquad (8.70)$$

The steady-state solutions (if they exist) of (8.68)–(8.70) can be obtained by employing Abel's Theorem to Laplace Transform, that is,

$$\lim_{s \to 0} sf(s) = \lim_{t \to \infty} f(t) \qquad (8.71)$$

8.5.5   A Repairable Markov Model of Two-Unit in Series II (Partial and Catastrophic Failure Modes)

Consider two three-state devices arranged in series [34]. The three-state device is repaired only when it fails in a partial mode (i.e., the other three-state device is operating successfully) or both devices are operating in their partial failure mode. Two automatic machines performing some operations on the assembly line items represent a typical example. The transition diagram for this series configuration is shown in Figure 8.13.

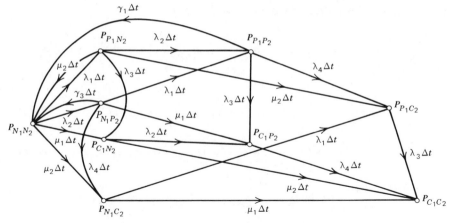

Figure 8.13 A series two-unit repairable Markov model.

Assumptions

1. Failures are statistically independent.
2. Repaired system is as good as new.
3. Failure and repair rates are constant.

Abbreviations and Notations

$P(t)$ = probability of state in question at time t.
$(\cdot)(\cdot)$

N_i = normal state of both three-state devices, $i = 1, 2$
C_1 = catastrophic failure state of the first three-state device
C_2 = catastrophic failure state of the second three-state device
P_1 = partial failure state of the first three-state device
P_2 = partial failure state of the second three-state device
λ_i = constant partial failure rates of both the devices, respectively, $i = 1, 2$
μ_i = constant catastrophic failure rates of both the devices, respectively, $i = 1, 2$
γ_i = constant system repair rates, $i = 1, 2, 3$
λ_3 = constant failure rate from partial to catastrophic failure mode of the first unit or device
λ_4 = constant failure rate from partial to catastrophic failure mode of the second unit
t = time
Δt = time interval
s = Laplace Transform variable

The differential equations associated with Figure 8.13 are

$$\frac{dP_{N_1N_2}(t)}{dt} + (\lambda_1 + \lambda_2 + \mu_1 + \mu_2)P_{N_1N_2}(t) = \gamma_1 P_{P_1P_2}(t) + \gamma_2 P_{P_1N_2}(t) + \gamma_3 P_{N_1P_2}$$

(8.72)

$$\frac{dP_{N_1P_2}(t)}{dt} + (\lambda_1 + \mu_1 + \gamma_3 + \lambda_4)P_{N_1P_2}(t) = P_{N_1N_2}(t)\lambda_2 \qquad (8.73)$$

$$\frac{dP_{P_1N_2}(t)}{dt} + (\lambda_2 + \mu_2 + \gamma_2 + \lambda_3)P_{P_1N_2}(t) = P_{N_1N_2}(t)\lambda_1 \qquad (8.74)$$

$$\frac{dP_{P_1P_2}(t)}{dt} + (\lambda_3 + \lambda_4 + \gamma_1)P_{P_1P_2}(t) = P_{P_1N_2}(t)\lambda_2 + P_{N_1P_2}(t)\lambda_1 \quad (8.75)$$

$P_{N_1N_2}(0) = 1$, at $t = 0$ other initial condition probabilities are zero.

The values of $P_{N_1N_2}(s)$, $P_{N_1P_2}(s)$, $P_{P_1N_2}(s)$, $P_{P_1P_2}(s)$ are obtained from the above differential equations:

$$P_{N_1N_2}(s) = \frac{-(\mu_1 + \lambda_1 + \gamma_3 + \lambda_4)(s + \mu_2 + \lambda_2 + \gamma_2 + \lambda_3)(s + \lambda_3 + \lambda_4 + \gamma_1)}{\Delta}$$

(8.76)

$$\Delta = \begin{vmatrix} (s+\lambda_1+\lambda_3+\mu_1+\mu_2) & -\gamma_1 & -\gamma_2 & \gamma_3 \\ -\lambda_2 & 0 & 0 & (s+\mu_1+\lambda_1+\gamma_3+\lambda_4) \\ -\lambda_1 & 0 & (s+\mu_2+\lambda_2+\gamma_2+\lambda_3) & 0 \\ 0 & (s+\lambda_3+\lambda_4+\gamma_1) & -\lambda_2 & -\lambda_1 \end{vmatrix}$$

$$P_{P_1P_2}(s) = \frac{-\{\lambda_2\lambda_1(s+\mu_2+\lambda_2+\gamma_2+\lambda_3)\} + (\mu_1+\lambda_1+\gamma_3+\lambda_4)\lambda_1\lambda_2}{\Delta}$$

(8.77)

$$P_{P_1N_2}(s) = \frac{-(s+\mu_1+\lambda_1+\gamma_3+\lambda_4)(s+\lambda_3+\lambda_4+\gamma_1)\lambda_1}{\Delta} \qquad (8.78)$$

$$P_{N_1P_2}(s) = \frac{-\lambda_2(s+\mu_2+\lambda_2+\gamma_2+\lambda_3)(s+\lambda_3+\lambda_4+\gamma_1)}{\Delta} \qquad (8.79)$$

8.5.6 A Two-Failure-Mode System with Cold Stand-By Units

Mathematical model [16] presents a system with two failure mode units and N stand-by units. The operational unit can be repaired at one of its failure modes. This may be regarded as a minor failure mode, in a case where the on-line failures can be repaired at the place of equipment installation or the unit repair time is less than the unit replacement time. When the unit repair is costly and time consuming, the failed unit is replaced with one of the stand-by units. Some of the typical examples of such a system may be the production line machinery, transformers, motors, heavy duty electrical switches, and so on.

Assumptions

1. It is assumed the repaired or replaced unit is as good as new.
2. The unit repair rate is faster than the unit replacement rate.
3. System fails only when the last standby unit fails.
4. The unit failed in its catastrophic mode is never repaired.
5. Failures are statistically independent.
6. A unit has two failure modes. Units can not fail in their standby mode.

Mathematical Model. The transition diagram of this system is shown in Figure 8.14. The following definitions and notations are used to formulate this mathematical model:

N = number of identical standby units
n = last state number of the system
μ_i = constant replacement and repair rates, respectively, of the operational unit for $i = 1, 2$ and $\mu_2 > \mu_1$
λ_1 = constant unit replacement mode failure rate
λ_2 = constant noncatastrophic mode (repairable mode) failure rate
t = time
s = Laplace transform variable
$P_0(t)$ = unit operational mode probability at time t
$P_1(t)$ = unit repairable mode probability at time t
$P_{k-i}(t)$ = unit failure, system operational and, system repairable mode probabilities at time t, for $i = 2, 1, 0$ respectively, and $k = 4, 7, 10, \ldots, (n-1)$
$P_n(t)$ = system failure mode probability at time t

The system differential equations for the Figure 8.14 model are

$$P_0'(t) = -(\lambda_1 + \lambda_2)P_0(t) + P_1(t)\mu_2 \tag{8.80}$$

$$P_1'(t) = \mu_2 P_1(t) + P_0(t)\lambda_1 \tag{8.81}$$

$$P_{k-2}'(t) = -\mu_1 P_{k-2}(t) + P_{k-4}(t)\lambda_2 \tag{8.82}$$

$$P_{k-1}'(t) = -(\lambda_1 + \lambda_2)P_{k-1}(t) + P_k(t)\mu_2 + P_{k-2}(t)\mu_1 \tag{8.83}$$

$$P_k'(t) = -\mu_2 P_k(t) + P_{k-1}(t)\lambda_1 \tag{8.84}$$

$$P_n'(t) = P_{k-1}(t)\lambda_2 \tag{8.85}$$

for $k = 4, 7, 10, \ldots, (n-1)$

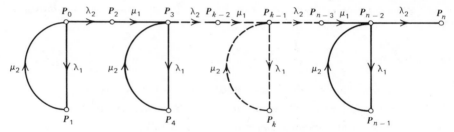

Figure 8.14 A transition diagram.

At $P_0(0) = 1$, other initial condition probabilities are equal to zero.

$$n = 3(N+1) - 1 \qquad \text{for} \quad N > 1 \tag{8.86}$$

where the prime denotes differentiation with respect to time t. The Laplace transforms of the solution are

$$P_0(s) = \frac{\{s + \mu_2\}}{\{(s + \mu_2)(s + \lambda_1 + \lambda_2) - \lambda_1 \mu_2\}} \tag{8.87}$$

$$P_1(s) = \frac{\{P_0(s)\lambda_1\}}{\{s + \mu_2\}} \tag{8.88}$$

$$P_{k-2}(s) = \frac{P_{k-4}(s)\lambda_2}{s + \mu_1} \tag{8.89}$$

$$P_{k-1}(s) = \frac{P_k(s)\mu_2 + P_{k-2}(s)\mu_1}{s + \lambda_1 + \lambda_2} \tag{8.90}$$

$$P_k(s) = \frac{P_{k-1}(s)\lambda_1}{s + \mu_2} \tag{8.91}$$

$$P_n(s) = \frac{P_{n-1}(s)\lambda_2}{s}. \tag{8.92}$$

8.5.7 Availability Analysis of a Two-Failure Modes System with Nonrepairable Stand-by Units

This model considers a system containing N identical units of which one is functioning and $(N-1)$ are standbys. As soon as the operational unit fails in any one of the two failure modes, it is replaced by one of the $(N-1)$ standby units. The system functions until the last standby unit is operational. The transition diagram of the Markov model is shown in Figure 8.15.

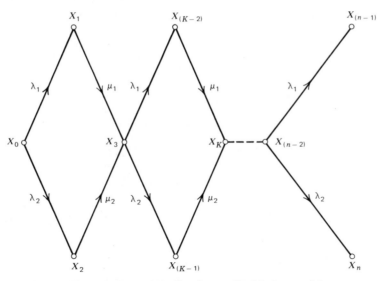

Figure 8.15 An $(N-1)$ units standby Markov model.

Notation

X_i = system operational (i.e., for $i=0,3,6,9,\ldots(n-2)$), failure mode type I (i.e., for $i=1,4,7,10,\ldots,(n-1)$) and failure mode type II (i.e., for $i=2,5,8,\ldots,n$) states

$P_i(t)$ = system operational (i.e., for $i=0,3,6,9,\ldots,(n-2)$), failure mode type I (i.e., for $i=1,4,7,10,\ldots,(n-1)$), and failure mode type II (i.e., for $i=2,5,8,\ldots,n$) probabilities at time t

λ_i = constant type I and type II failure mode failure rates of the operational unit, respectively (i.e., for $i=1,2$)

μ_i = constant type I and type II failure mode state replacement rates of the failed unit, respectively (i.e., for $i=1,2$)

n = number of system states

N = number of units in the system (i.e., the operational unit plus standby units)

t = time

s = Laplace transform variable

Assumptions

1. Failures are statistically independent.
2. Restored unit is as good as new.
3. Cold standby units cannot fail.
4. Failed unit is never repaired.

The system of differential equations associated with Figure 8.15 are

$$P_0'(t) = -\sum_{i=1}^{2} \lambda_i P_0(t) \tag{8.93}$$

$$P_{(K-1)}'(t) = P_{(K-3)}(t)\lambda_2 - P_{(K-1)}(t)\mu_2 \tag{8.94}$$

$$\text{for} \quad K=3,6,9,12,\ldots,(n-2)$$

$$P_{(K-2)}'(t) = P_{(K-3)}(t)\lambda_1 - P_{(K-2)}(t)\mu_1 \tag{8.95}$$

$$\text{for} \quad K=3,6,9,12,\ldots,(n-2)$$

$$P_K'(t) = P_{(K-2)}(t)\mu_1 + P_{(K-1)}(t)\mu_2 - \sum_{i=1}^{2} \lambda_i P_K(t) \tag{8.96}$$

$$\text{for} \quad K=3,6,9,12,\cdots(n-2)$$

$$P_{(n-1)}'(t) = P_{(n-2)}(t)\lambda_1 \tag{8.97}$$

$$P_n'(t) = P_{(n-2)}(t)\lambda_2 \tag{8.98}$$

At $P_0(0)=1$, other initial condition probabilities are zero.

$$n=(3N-1) \quad \text{for } N \geqslant 1 \tag{8.99}$$

where the prime denotes the derivative with respect to time t. Solutions to the above system of differential equations in the s domain are

$$P_0(s) = \frac{1}{s + \sum\limits_{i=1}^{2} \lambda_i} \tag{8.100}$$

$$P_{(K-1)}(s) = \frac{P_{(K-3)}(s)\lambda_2}{s+\mu_2} \tag{8.101}$$

$$\text{for} \quad K=3,6,9,12,\ldots,(n-2)$$

$$P_{(K-2)}(s) = \frac{P_{(K-3)}(s)\lambda_1}{s+\mu_1} \tag{8.102}$$

$$P_K(s) = \left\{ P_{(K-2)}(s)\mu_1 + P_{(K-1)}(s)\mu_2 \right\} \frac{1}{s + \sum\limits_{i=1}^{2} \lambda_i} \tag{8.103}$$

for $K = 3, 6, 9, 12, \ldots, (n-2)$

$$P_{(n-1)}(s) = \frac{P_{(n-2)}(s)\lambda_1}{s} \qquad (8.104)$$

$$P_n(s) = \frac{P_{(n-2)}(s)\lambda_2}{s} \qquad (8.105)$$

To obtain state probabilities invert (8.100)–(8.105) to time domain [i.e., take inverse Laplace transforms of (8.100)–(8.105)]. The system operational availability, A_s, can be obtained from

$$A_s = \sum_{i=0}^{n-2} P_i(t) \qquad (8.106)$$

for $i = 0, 3, 6, 9, \ldots, (n-2)$

8.5.8 A k-*out-of*-n Three-State Device System with Common-Cause Failures

This section presents a generalized Markov model to represent repairable k-out-of-n units system with common-cause failures [14]. This mathematical model can also be applied to represent repairable series or parallel (two- or three-state device) network subject to common-cause failures. Some of the common-cause failures may occur due to (a) undetected design errors; (b) operator and maintenance errors; (c) common environments; (d) common manufacturer; (e) common energy source; (f) same repairman; or (g) equipment failure event—fire, flood, tornado, earthquake. A typical example may be a redundant configuration composed of two motorized fluid flow valves with common (control circuit) power supply. This type of situation is frequently encountered in power stations.

Assumptions

1. Three-state devices are identical.
2. The redundant system is only repaired when all devices fail in either failure modes (i.e., open, short, closed), or if the redundant system fails due to common-cause failures.
3. Common-cause failures can only occur if two or more three-state devices are present in a system.
4. A failed system is restored as good as new.
5. Common-cause and other failures are statistically independent.

Notation

λ_i = constant open mode failure rate, for $i = 0, 1, 2, 3, \ldots, k$

α_i = constant short mode failure rate, for $i = 0, 1, 2, 3, \ldots, k$

γ_i = constant common-cause failure rate, for $i = 0, 1, 2, 3, \ldots, (k-1)$

μ_{SH} = constant short failure mode repair rate

μ_o = constant open failure mode repair rate

μ_c = constant common-cause failure mode repair rate

$P_i(t)$ = state probability at time t for $i = 0, 1, 2, 3, \ldots, n$

  (Note: for $i = n$ represents open failure mode probability at time t)

$P_c(t)$ = common-cause failure mode probability at time t

$P_{SH}(t)$ = short failure mode probability at time t

N = total number of devices in a system

s = Laplace transform variable

t = time

The associated equations with Figure 8.16 are

$$P_0'(t) = -(\lambda_0 + \alpha_0 + \gamma_0)P_0(t) + P_{SH}(t)\mu_{SH} + P_c(t)\mu_c + P_n(t)\mu_o$$

$$(8.107)$$

$$P_1'(t) = -(\lambda_1 + \alpha_1 + \gamma_1)P_1(t) + P_0(t)\lambda_0 \qquad (8.108)$$

$$P_2'(t) = -(\lambda_2 + \alpha_2 + \gamma_2)P_2(t) + P_1(t)\lambda_1 \qquad (8.109)$$

$$\vdots$$

$$P_{k-1}'(t) = -(\lambda_{k-1} + \alpha_{k-1} + \gamma_{k-1})P_{k-1}(t) + P_{k-2}(t)\lambda_{k-2} \qquad (8.110)$$

$$\vdots \quad \text{for} \quad k = 2, 3, 4, \ldots, (n-1),$$

$$P_k'(t) = -(\lambda_k + \alpha_k)P_k(t) + P_{k-1}(t)\lambda_{k-1} \qquad (8.111)$$

$$\vdots \quad \text{for} \quad k = (n-1),$$

$$P_n'(t) = -\mu_o P_n(t) + P_k(t)\lambda_k \qquad (8.112)$$

$$P_{SH}'(t) = -\mu_{SH}P_{SH}(t) + \sum_{i=0}^{k} \alpha_i P_i(t) \qquad \text{for} \quad k = n-1, \quad (8.113)$$

$$P_c(t) = -\mu_c P_c(t) + \sum_{i=0}^{k-1} \gamma_i P_i(t) \qquad \text{for} \quad k = n-1 \qquad (8.114)$$

$$n = N \qquad \text{for} \quad N \geqslant 2$$

$$\lambda_i = (N-i)\lambda \qquad \text{for} \quad i = 0, 1, 2, 3, \ldots, N$$

At $P_0(0) = 1$, other initial condition probabilities are equal to zero.

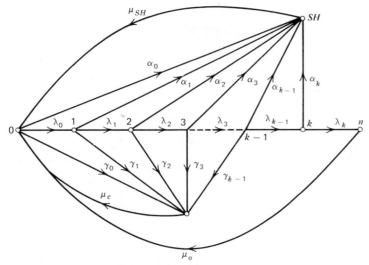

Figure 8.16 Transition diagram.

The prime denotes differentiation with respect to time t. Laplace transforms of the state probability equations are

$$P_0(s) = \frac{1 + P_{SH}(s)\mu_{SH} + P_c(s)\mu_c + P_n(s)\mu_o}{s + \lambda_0 + \alpha_0 + \gamma_0} \tag{8.115}$$

$$P_1(s) = \frac{P_0(s)\lambda_0}{s + \lambda_1 + \alpha_1 + \gamma_1} \tag{8.116}$$

$$P_2(s) = \frac{P_1(s)\lambda_1}{s + \lambda_2 + \alpha_2 + \gamma_2} \tag{8.117}$$

$$\vdots$$

$$P_{k-1}(s) = \frac{P_{k-2}(s)\lambda_{k-2}}{s + \lambda_{k-1} + \alpha_{k-1} + \gamma_{k-1}} \tag{8.118}$$

$$\vdots \quad \text{for} \quad k = 2, 3, 4, \ldots, (n-1),$$

$$P_k(s) = \frac{P_{k-1}\lambda_{k-1}}{s + \lambda_k + \alpha_k} \quad \text{for} \quad k = (n-1), \tag{8.119}$$

$$\vdots$$

$$P_n(s) = \frac{P_k(s)\lambda_k}{s + \mu_o} \tag{8.120}$$

$$P_{SH}(s) = \frac{\displaystyle\sum_{i=0}^{k} \alpha_i P_i(s)}{s + \mu_{SH}} \quad \text{for} \quad k = n-1 \tag{8.121}$$

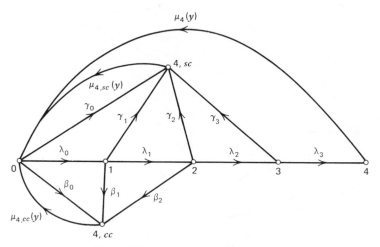

Figure 8.17   A system transition diagram.

$$P_c(s) = \frac{\displaystyle\sum_{i=0}^{k-1} \gamma_i P_i(s)}{s + \mu_c} \qquad \text{for} \quad k = n-1. \qquad (8.122)$$

To use this model for series configuration interchange open failure mode probability with short (close) failure mode probability.

8.5.9   A 4-Unit Redundant System with Common-Cause Failures

This model [13] can be used for devices with two mutually exclusive failure modes and common-cause failures. The transition diagram of the model is shown in Figure 8.17.

Assumptions

1. Common-cause and other failures are s-independent.
2. Common-cause failures can only occur with more than one unit. 4-units are identical.
3. Units are repaired only when the system fails. A failed system is restored as good as new.
4. System repair times are arbitrarily distributed.

The transition diagram is shown in Figure 8.17.

Notation

$i =$ state of the unfailed system; number of failed units; $i = 0, 1, 2, 3$

$j=$ state of the failed system; $j=4$ means failure not due to a common cause; $j=4,cc$ means failure due to a common-cause; $j=4,sc$ means short (closed) mode of failure

$P_i(t)=$ probability that system is in unfailed state i at time t

$P_j(y,t)=$ probability density (with respect to repair time) that the failed system is in state j and has an elapsed repair time of y

$\mu_j(y), q_j(y)=$ repair rate (a hazard rate) and probability density function of repair time when system is in state j and has an elapsed repair time of y

$\beta_i=$ constant common-cause failure rate of the system when in state i; $\beta_3=0$

$\lambda_i=$ constant failure rate of a unit, for other than common-cause failures, when the system is in state i; $i=0,1,2,3$

$s=$ Laplace transform variable

$\gamma_i=$ constant short (closed mode) failure rate when the system is in state $i; 0,1,2,3$

Equations (8.123)–(8.128) associated with Figure 8.17 are

$$\frac{dP_0(t)}{dt}+(\lambda_0+\beta_0+\gamma_0)P_0(t)=\int_0^\infty P_{4,cc}(y,t)\mu_{4,cc}(y)\,dy$$

$$+\int_0^\infty P_4(y,t)\mu_4(y)\,dy+\int_0^\infty P_{4,sc}(y,t)\mu_{4,sc}(y)\,dy$$

$$(8.123)$$

$$\frac{dP_i(t)}{dt}+(\lambda_i+\beta_i+\gamma_i)P_i(t)-\lambda_{i-1}P_{i-1}(t)=0 \qquad (8.124)$$

$$\text{for}\quad i=1,2,3;\quad \beta_3=0$$

$$\frac{\partial p_j(y,t)}{\partial t}+\frac{\partial p_j(y,t)}{\partial y}+\mu_j(y)p(y,t)=0 \qquad (8.125)$$

$$p_4(o,t)=\lambda_3 P_3(t) \qquad (8.126)$$

$$p_{4,cc}(o,t)=P_0(t)\beta_0+P_1(t)\beta_1+P_2(t)\beta_2 \qquad (8.127)$$

$$p_{4,sc}(o,t)=P_0(t)\beta_0+P_1(t)\gamma_1+P_2(t)\gamma_2+P_3(t)\gamma_3 \qquad (8.128)$$

$$\lambda_i=(i-1)\lambda$$

$$\gamma_i=(i-2)\gamma$$

$$P_i(o)=1 \qquad \text{for}\quad i=0\quad \text{other}\ P_i(o)=0$$

$$P_j(y,0)=0 \qquad \text{for all}\ j$$

The Laplace transforms of the solution of (8.123)–(8.128)

$$P_0(s) = \left[s + \lambda_0 + \beta_0 + \gamma_0 - \left(\beta_0 + \frac{\beta_1}{A_1} + \frac{\beta_2}{A_2} \right) G_{4,cc}(s) \right.$$

$$\left. - \frac{G_4(s)\lambda_3}{A_3} - \left(\gamma_0 + \frac{\gamma_1}{A_1} + \frac{\gamma_2}{A_2} + \frac{\gamma_3}{A_3} \right) G_{4,sc} \right]^{-1} \qquad (8.129)$$

$$G_j(s) \equiv \int_0^\infty \exp(-sy) q_j(y) \, dy \qquad \text{for } j = 4, 4, cc, \text{ or } 4, sc$$

$$A_1 \equiv \frac{s + \lambda_1 + \beta_1 + \gamma_1}{\lambda_0}$$

$$A_2 \equiv \frac{A_1(s + \lambda_2 + \beta_2 + \gamma_2)}{\lambda_1} \qquad A_3 \equiv \frac{A_2(s + \lambda_3 + \gamma_3)}{\lambda_2}$$

$$P_i(s) = \frac{P_0(s)}{A_i} \qquad \text{for } i = 1, 2, 3 \qquad (8.130)$$

$$P_4(s) = \lambda_3 P_3(s) \frac{1 - G_4(s)}{s} \qquad (8.131)$$

$$P_{4,cc}(s) = \left[\sum_{i=0}^2 \beta_i P_i(s) \right] \frac{1 - G_{4,cc}(s)}{s} \qquad (8.132)$$

$$P_{4,sc}(s) = \left[\sum_{i=0}^3 \gamma_i P_i(s) \right] \frac{1 - G_{4,sc}(s)}{s} \qquad (8.133)$$

To obtain time domain solutions, (8.129)–(8.133) can be transformed for given repair times distribution.

8.6 RELIABILITY OPTIMIZATION OF THREE-STATE DEVICE NETWORKS

This section deals with optimizing the number of redundant elements to obtain maximum reliability. Here, we focus on obtaining the optimum number of redundant elements for the series and parallel configuration only.

8.6.1 Series Network

Using expression 8.1 the series system reliability of identical elements is given by

$$R = (1-q_0)^n - q_s^n \tag{8.134}$$

To obtain optimum number of elements differentiate (8.134) with respect to n and equate it to zero. The following results are obtained

$$\frac{\partial R}{\partial n} = \alpha_0^n \log_e \alpha_0 - q_s^n \log_e q_s = 0 \tag{8.135}$$

where $\alpha_0 = (1-q_0)$.

Thus, rewriting (8.135) in terms of n optimum number of elements n^*, we get

$$n^* = \frac{\log_e \left\{ \dfrac{\log_e q_s}{\log_e \alpha_0} \right\}}{\log_e (\alpha_0 / q_s)} \tag{8.136}$$

8.6.2 Parallel Network

The following expression is directly obtained from (8.136) by reasoning the duality of the series to parallel form

$$m^* = \frac{\left[\log_e (\log_e q_0 / \log_e \alpha_s) \right]}{\log_e (\alpha_s / q_0)} \tag{8.137}$$

where $\alpha_s = (1-q_s)$ and m^* is the optimum number of elements. Optimization of series or parallel network reliability subject to constraints is presented in reference 24.

REFERENCES

Three-State Devices

1. Barlow, R. E. and L. C. Hunter, "Criteria for Determining Optimum Redundancy," *Trans. IRE Reliab. Qual. Control*, **9**, 73–77 (1960).
2. Barlow, R. E. and L. C. Hunter, "Mathematical Models for System Reliability, Part II," *Sylvania Technol.*, **13**, 55–65 (1960).
3. Barlow, R. E., L. C. Hunter, and F. Proschan, "Optimum Redundancy when Components Subject to Two Kinds of Failure," *J. Indust. Appl. Math.* II (1963), pp. 64–73.
4. Blake, D. V., "Open and Short Circuit Failure of Hammock Networks," *Electron. Reliab. Microminiat.*, **2**, 205–206 (1963).

5. Butler, B. E., "A Review of Present Practice in Reliability Engineering," *Microelectron. Reliab.*, **10**, 195–208 (1971).

6. Cluley, J. C., "Low Level Redundancy as a Means of Improving Digital Computer Reliability," *Electron. Reliab. Microminiat.*, **1**, 203–216 (1962).

7. Creveling, C. J., "Increasing the Reliability of Electronic Equipment by the Use of Redundant Circuits," *Proc. IRE*, **44**, 509–515 (1956).

8. Dhillon, B. S., "The Analysis of the Reliability of Multi-State Device Networks," PhD Dissertation, National Library of Canada, Ottawa.

9. Dhillon, B. S., "Analysis of a Three-State System with Two Types of Components," *Microelectron. Reliab.*, **15**, 245–246 (1976).

10. Dhillon, B. S., "Reliability Evaluation of Three-State Device Network Models—An Electrical Analogous Technique," *Proc. 7th Modelling and Simulation Conference*, Pittsburgh, Instrument Society of America, Pittsburgh, 1976, pp. 225–229.

11. Dhillon, B. S., "Availability Models for a System of *N*-Redundant Three-State Equipment," *Proceedings on 1977 Environmental Technology*, Institute of Environmental Sciences, Mt. Prospect, IL, pp. 235–237.

12. Dhillon, B. S., "Literature Survey on Three-State Device Reliability Systems," *Microelectron. Reliab.*, **16**, 601–602 (1977).

13. Dhillon, B. S., "A 4-Unit System with Common-Cause Failures," *IEEE Trans. Reliab.*, **R-29**, pp. 267 (1979).

14. Dhillon, B. S., "A *k*-out-of-*N* Three-State Device System with Common-Cause Failures," *Microelectron. Reliability*, **19** (1978).

15. Dhillon, B. S. and A. Sambhi, "Hazard Rate Comparison of Multi-State Device Systems," *Proceedings of the International JSME Symposium*, Japan Society of Mechanical Engineers, Tokyo, Japan, 1977.

16. Dhillon, B. S., "A Two Failure Modes System with Cold Stand-By Units," *Microelectron. Reliab.*, **19** (1978).

17. Dhillon, B. S., et. al., "Common-Cause Failure Analysis of a Three-State Device System," *Microelectron. Reliability* **19** (1979).

18. Dubes, R. C., "Two Algorithms for Computing Reliability," *IEEE Trans. Reliab.*, **R-12**, 55–63 (1963).

19. Evans, R. A., "Redundancy Configurations and Their Effects on System Reliability," *Microelectron. Reliab.* **10**, 355–357 (1971).

20. Gopal, K. et al., "Reliability Analysis of Multistate Device Networks," *IEEE Trans. Reliab.* **R-27**, 233–236 (1978).

21. Henin, C., "Double Failure and Other Related Problems in Standby Redundancy," *IEEE Trans. Reliab.*, **R-21**, 35–40 (1972).

22. Hyun, K. N., "Reliability Optimization by 0-1 Programming for a System with Several Failure Modes," *IEEE Trans. Reliab.*, **R-24**, 206–210 (1975).

23. James, D. C., A. H. Kent, and J. A. Holloway, "Redundancy and Detection of First Failures," *IRE Trans. Reliab. Qual. Control*, **11**, 8–28 (1962).

24. Kolesar, P. J., "Linear Programming and the Reliability of Multicomponent Systems," *Naval Res. Logist. Quart.*, **14**, 317–328 (1967).

25. Lipp, J. P., "Topology of Switching Elements Vs. Reliability," *IRE Trans. Reliability Qual. Contr.*, **6**, 35–39 (1957).

26. Misra, K. B. and T. S. M. Rao, "Reliability Analysis of Redundant Networks Using Flow Graphs," *IEEE Trans. Reliability*, **19**, 19–24 (1971).

27. Moore, E. F. and C. E. Shannon, "Reliable Circuits Using Less Reliable Relays," *J. Franklin Inst.* **9**, 191–208 and **10**, 1–297 (1956).

28. Premo, A. F., "The Use of Boolean Algebra and a Truth Table in the Formulation of a Mathematical Model of Success," *IEEE Trans. Reliab.* **R-12**, 45–49 (1963).

29. Price, W. P., "Reliability of Parallel Electronic Components," *IRE Trans. Reliability Qual. Contr.*, **9**, 35–39 (1960).

30. Proctor, C. L. and B. Singh (Dhillon), "A Repairable Three-State Device," *IEEE Trans. Reliab.*, **25**, 210–211 (1976).

31. Proctor, C. L. and B. Singh (Dhillon), "Reliability of Three-State Device Networks," *Proceedings of the Annual Reliability and Maintainability Symposium*, IEEE, New York, 1975, pp. 311–316.

32. Singh (Dhillon), B. and C. L. Proctor, "Reliability Analysis of Multistate Device Networks," *Proceedings of the Annual Reliability and Maintainability Symposium*, IEEE, New York, 1976, pp. 31–35.

33. Singh (Dhillon), B. and C. L. Proctor, "A Parametric Technique to Evaluate Reliability of Three-State Device Networks," *Proceedings of the Conference on System Science*, University of Roorke, U.P., India, 1976.

34. Singh (Dhillon), B. and C. L. Proctor, "Reliability Analysis of Repairable Three-State Device Markov Models," *Proceedings of the Conference on Applied Reliability*, University of South Florida, 1975.

35. Sorensen, A. A., "Digital-Circuit Reliability through Redundancy," *Electron. Reliability Microminiat.*, **1**, 203–216 (1962).

36. Tillman, F. A., "Optimization by Integer Programming of Constrained Reliability Problems with Several Modes of Failure," *IEEE Trans. Reliability*, **18**, 47–53 (1969).

37. Webster, L., "Choosing Optimum System Configurations," *Proceedings of the Annual Reliability Symposium*, IEEE, New York, 1964, pp. 345–358.

38. Wiggins, A. D., "Reliability of Four-State Safety Devices," *Microelectron. Reliability*, **11**, 335–353 (1972).

Miscellaneous

39. Cox, D. R., "The Analysis of Non-Markovian Stochastic Processes by the Inclusion of Supplementary Variables," *Proc. Camb. Phil. Soc.*, **61**, 433–41 (1955).

40. Garg, R. C., "Dependability of a Complex System Having Two Types of Components," *IEEE Trans. Reliab.*, **R-12**, 11–15 (1963).

41. Gaver, D. P., "Time to Failure and Availability of Paralleled Systems with Repair," *IEEE Trans. Reliability*, **R-12**, 30–38 (1963).

9

Power System Reliability

9.1 INTRODUCTION

A primary requirement of a modern electric power system is a reasonable ability to satisfy the customer load requirements. In some electric utilities this involves generation, transmission, and distribution facilities. In others, the responsibility may extend over a part of the total facility. A complete power system is, however, composed of generation, transmission, and distribution facilities each one of which contributes its own inherent difficulties to the problem of satisfying customer requirement. A power system should be designed and expansion facilities planned so that it can perform its intended function with a reasonable risk. The risk of power interruption or capacity shortage can be reduced by providing more redundancy in the transmission and distribution networks and enough reserve generating capacity. There has, however, to be a trade off between reliability of power supply and the cost involved. Reliability models provide a means of carrying out this trade off.

There has been considerable growth in the techniques for the quantitative evaluation of the reliability of power systems. Because of the structural similarity of the various power systems, a number of generic reliability techniques have been developed for planning, design, and operation of these systems. This ensemble of concepts, indices, and methods is generally referred to as the power system reliability. Numerous papers [12] and two books [9, 11] have been written on this subject. This chapter gives a compact and unified approach to power system reliability and references to more detailed discussions are provided.

The major areas of a power system are generation, transmission, and distribution. For determining the reliability indices, the entire power system is not considered. Although conceptually possible, the complexity and dimensionality make this task rather impractical at present. It appears, however, that this task will become possible in the future partly by developments in better techniques of modeling and partly due to an increase in both the speed and power of computers. At present, however, the major divisions in power system reliability are the generating capacity reliability, bulk power system reliability, interconnected systems reliability, and the reliability of distributions systems.

9.2 GENERATING CAPACITY RELIABILITY

Generating capacity reliability evaluation can be considered in two basic forms, which may be designated static reserve and operating reserve requirements. The static reserve studies are concerned with determining the installed reserve capacity sufficient to provide for unplanned and planned outages of generating units and uncertainties in the forecast load. The operating reserve consists of spinning or quick starting units and is a capacity that must be available to meet load changes and also capable of satisfying the loss of some portion of generating capacity. Whereas the static reserve is of primary concern to the planning engineer, the operating reserve provides assistance in decisions on daily operation of the power system. Ideally both of these areas must be investigated at planning level but once a decision has been reached, the operating reserve becomes an operating problem. This section is concerned with the static reserve area and the operating reserve is described in Section 3.

Generating capacity reliability studies assume the transmission network to be perfectly reliable and capable of transferring the energy from any generation point to the load point. This amounts to the assumption that all the generating units and loads are connected across a single bus. The assessment is basically concerned with the certainty with which the system load can be satisfied by the generation facilities. The three basic steps involved are [17]:

1. A model describing the probabilistic behaviour of capacity outages is developed first. This is referred to as the "generation system model."
2. The probabilistic nature of the daily load curve is incorporated into a "demand model" or "load model."
3. The generation and load models are then merged or convolved to give a "generation reserve model," which depicts the expected occurrence of surplus capacity and capacity deficiencies. Several indices are defined on the generation reserve as measures of generating capacity reliability.

9.2.1 Generation System Model

Model of a Single Unit. A generating unit, especially a large thermal one, may have several different capacity levels. The consideration of partial or derated capacity states is not a major problem; however, in order to illustrate the basic approach each unit is assumed to exist either in an up (full capacity) or in a down (zero capacity) state. This binary model can be characterized by the following parameters:

c = capacity of the unit in MW
m = mean up time of the unit
r = mean down time of the unit

Using these parameters, the model of the single unit can be expressed as follows:

$$\Pr(\text{capacity out}=0) = \frac{m}{m+r}$$

$$= \frac{\mu}{\lambda+\mu} \tag{9.1}$$

$$\Pr(\text{capacity out}=c) = \frac{r}{m+r}$$

$$= \frac{\lambda}{\lambda+\mu} \tag{9.2}$$

$$\Fr(\text{capacity out}=0) = \Fr(\text{capacity out}=c)$$

$$= \frac{\lambda\mu}{\lambda+\mu} \tag{9.3}$$

where

$\lambda, \mu =$ the reciprocals of m and r and are called the failure and repair rates of the unit

$\Pr(\cdot), \Fr(\cdot) =$ the steady-state probability and frequency of $(\cdot)$, respectively.

Equations 9.1–9.3 are independent of the form of the probability density function of the up and down times. Equation 9.2 can be recognized as the unavailability of the unit and has been traditionally called "forced outage rate" and defined as [13]

$$\text{FOR} = \frac{\text{forced outage hours}}{\text{in-service hours}+\text{forced outage hours}} \tag{9.4}$$

System Model. The generation system consists of many units and they are assumed statistically independent. Such a system can have many possible capacity levels and the reliability measures of the following form are required.

1. *Pr*(lost system capacity $\geqslant x$), the steady-state probability of lost capacity equal to or greater than x MW. The state, $\geqslant x$ is commonly called cumulative state as compared with the exact state, that is, equal to x MW.
2. *Fr*(lost system capacity $\geqslant x$), the mean frequency of encountering the cumulative state of x or more MW lost capacity. The reciprocal of this function gives the mean cycle time, that is, the mean time between two successive encounters of this state.

The problem for a small system can be solved by enumerating the possible system states, calculating the probabilities by the multiplication rule, and then determining the subset probabilities by addition. The frequencies can be readily calculated using the frequency balancing approach [19]. This method is practicable, however, only as long as the number of units comprising the system is relatively small. A generation system may have several hundred units and therefore this approach is not feasible. The generation system model is typically developed by the sequential addition of units.

Algorithm for Unit Addition. This algorithm [20] is basic to system model building by sequential unit addition. The system is assumed to consist of binary units, that is, it can exist either in an up (full capability) or down (zero capability) state. There is, however, no inherent limitation in developing similar algorithms for multistate units using the same methodology. The assumption of binary units is for keeping the discussion simple. It has been already shown in reference 26 that the probability density functions of up and down times do not effect the steady-state probabilities and mean frequencies when the units are assumed statistically independent. The following notation is used:

$\overline{C}_i = i$th lost capacity state. Capacity states are assumed to be arranged in an increasing order of lost capacity, $\overline{C}_{i+1} > \overline{C}_i$

$P_i, f_i =$ steady-state probability and mean frequency of lost capacity $\geqslant \overline{C}_i$

$c_k =$ capacity of the kth unit

$\lambda_k = 1/m_k$

$m_k =$ mean up time of unit k

$\mu_k = 1/r_k$

$p_k = \mu_k/(\lambda_k + \mu_k),$

$\left.\begin{matrix} P_{(i+k)} \\ f_{(i+k)} \end{matrix}\right\} =$ steady-state probability and mean frequency of lost capacity $\geqslant (C_i + c_k)$

$N_k =$ number of capacity states for the k unit system

$r_k =$ mean down time of unit k

$a =$ to "after," that is, after the unit addition or after the unit removal (superscript)

The process of model building is started with a single unit and then each unit is added in turn. The model for a single unit is, of course, quite straightforward. Now assume that a system model exists for $(k-1)$ units and it is required to add the kth unit. Since the unit is assumed to exist either in the up state (lost capacity $= 0$) or in the down state (lost capacity $= c_k$), two groups of system states would be obtained after unit addition, $\{\overline{C}_i + 0\}$ and $\{\overline{C}_i + c_k\}$, $i = 1, 2, \ldots, N$, (see Figure 9.1). The states in the

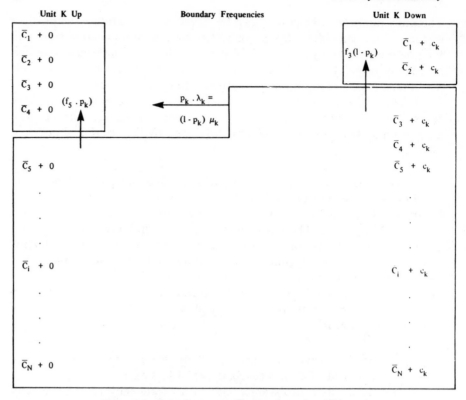

Figure 9.1 State frequency diagram for unit addition.

former group are termed "existing lost capacity states" and those in the latter "generated lost capacity states." As an example consider $\bar{C}_5$ such that $(\bar{C}_3 + c_k) \geqslant \bar{C}_5$ and $(\bar{C}_2 + c_k) < \bar{C}_5$. The boundary for this state is shown in Figure 9.1 and the associated frequency is

$$= f_5 p_k + f_3(1 - p_k) + (P_3 - P_5)p_k\lambda_k \qquad (9.5)$$

In general, modified cumulative probability and frequency of existing lost capacity state i is given by

$$P_i^m = P_i^o p_k + P_j^o(1 - p_k) \qquad (9.6)$$

and

$$f_i^m = f_i^o \cdot p_k + f_j^o(1 - p_k) + (P_j^o - P_i^o)p_k \cdot \lambda_k, \qquad (9.7)$$

such that

$$\bar{C}_j \geqslant (\bar{C}_i - c_k)$$

Superscript o refers to the old values of probabilities and frequencies and superscript m refers to the modified values after unit addition. The probabilities and frequencies of generated lost capacity states are similarly given by

$$P_{(i+k)} = P_i^o(1-p_k) + P_j^o \cdot p_k \tag{9.8}$$

and

$$f_{(i+k)} = f_i^o(1-p_k) + f_j^o p_k + (P_i^o - P_j^o)p_k\lambda_k \tag{9.9}$$

such that

$$\overline{C}_j \geqslant (\overline{C}_i + c_k)$$

The flow diagram for the computer implementation of this algorithm is provided in reference 20. The procedure described in reference 20 accomplishes calculation and state reordering at the same time and is very fast.

Algorithm for Unit Removal. Several times it may be necessary to remove a unit from the system model. For example, during the period of a year different units are on scheduled maintenance and therefore the same system model cannot be used for the whole period. The year can be divided into a number of intervals during which the units on scheduled maintenance stay the same and a single system model can be used. These system models can be derived from the master system model by removing the units on maintenance.

Unit removal is the reverse of the process of unit addition described by (9.6)–(9.9). To reconstruct the system model prior to the addition of unit k, (9.8) and (9.9) can be modified. Since $\overline{C}_j$ is equal to or just greater than $\overline{C}_i + c_k$,

$$P_{(i+k)}^o = P_j^o \tag{9.10}$$

and

$$f_{(i+k)}^o = f_j^o \tag{9.11}$$

Substituting (9.10) and (9.11) into (9.8) and (9.9) and replacing subscripts $(i+k)$ and i by i and j, respectively,

$$f_i^m = f_i^o \cdot p_k + f_j^o(1-p_k) + (P_j^o - P_i^o)p_k\lambda_k \tag{9.12}$$

and

$$P_i^m = P_i^o p_k + P_j^o \cdot (1-p_k) \tag{9.13}$$

such that

$$\bar{C}_j \geqslant (\bar{C}_i - c_k)$$

Equations 9.12 and 9.13 are the same as (9.6) and (9.7). Probabilities and frequencies of the old system model are therefore,

$$P_i^o = \frac{\left[P_i^m - P_j^o(1-p_k) \right]}{p_k} \qquad (9.14)$$

and

$$f_i^o = \frac{\left[f_i^m - f_j^o(1-p_k) + (P_i^o - P_j^o)p_k\lambda_k \right]}{p_k} \qquad (9.15)$$

such that

$$\bar{C}_j \geqslant \bar{C}_i - c_k$$

The algorithm is started with $P_1^o = 1$ and $f_1^o = 0$. After each unit removal, the system model may contain sets of states with different lost capacity values but the same probabilities and frequencies. In these sets, all the states except the last one should be deleted to obtain the exact system model prior to the addition of unit k. The flow diagram for the computer implementation of this algorithm is given in reference 20.

9.2.2   Load and Generation Reserve

Load Model for Loss of Load Expectation.   One of the indices used in static reserve studies is loss of load probability (LOLP) or loss of load expectation (LOLE). The load model generally employed for LOLE calculation is of the shape shown in Figure 9.2. This cumulative load curve indicates the time for which the load is more than a specified level in MW. This curve is either from hourly load durations or more generally from the daily peaks. In the latter case, it indicates the number of days on which the peak exceeded a specified value. The LOLE is an expected value and is given by [6].

$$\text{LOLE} = \sum p_i t_i \qquad (9.16)$$

where p_i = probability of a capacity outage equal to c_i
t_i = number of time units, in the study period, that a capacity outage of c_i would result in a loss of load

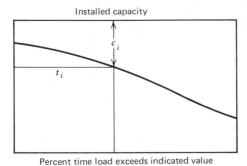

Installed capacity

c_i

t_i

Percent time load exceeds indicated value

Figure 9.2 Cumulative load duration curve.

The values of p_i can be obtained from the generation system model discussed in Section 9.2.1 by the following relationship,

$$p_i = P_i - P_{i+1}$$

The LOLE is expressed in days/year, that is, the expected time in days that the load would not be met by the capacity in a period of 1 year. The magnitude of load loss is not considered. The reciprocal of LOLE in years/day is often used as a reliability measure; however, it does tend to obscure the fact that LOLE is a simple expectation.

Load Model for Frequency and Duration. The load model for loss of load probability method does not adequately reflect the shape of the daily load variation curve. Reference 16 suggested a load model introducing an exposure factor to indicate that the peak load does not persist for the entire day. The mean duration at a particular load level, usually about 80% of the daily peak, is assumed to be exposure factor. This amounts to approximating the daily load variation curve as shown in Figure 9.3, and taking the mean of e_i as the exposure factor. The load model, then is assumed to consist of a random sequence of N load states, each of which is followed by a low load state (see Figure 9.4). The state transition diagram for this load model is shown in Figure 9.5 and some parameters are given below.

Description of load levels, MW	$L_i, i = 1, 2, \ldots, N$
	$L_1 > L_2 > \cdots > L_n$
Number of occurrences of L_i	n_i
Interval length	$D = \sum\limits_{i=1}^{N} n_i$
Exposure factor, that is, the mean duration of L_i	$e < 1$ days

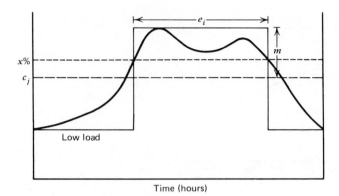

Figure 9.3   The daily load variation curve.

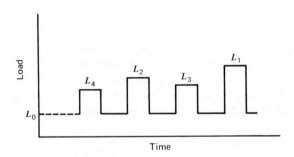

Figure 9.4   The basic load model.

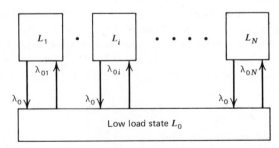

Figure 9.5   State transition diagram of the basic load model.

CAPACITY RESERVE MODEL. Assuming stochastic independence of generation system model and load model, they may be combined or convolved to form a capacity reserve model. Capacity reserve or margin is an excess of available capacity over demand, that is

$$\text{margin} = \text{capacity} - \text{load}$$

The probabilities and frequencies of cumulative margins (margin $\leq M$) are of primary interest and these may be computed by combining the cumulative capacity states with the exact load states. Using the margin state matrix approach [18], it can be proved that

$$A_M = \sum_{i=1}^{N+1} P_v A_i \tag{9.17}$$

where
$$\begin{aligned} A_M &= \text{steady-state probability of margin} \leq M \\ A_i &= n_i e / D \\ P_v &= \text{probability of capacity} \leq C_v \text{ such that } C_v \leq (L_i + M) \\ N+1 &\equiv \text{the low load state, } L_0 \end{aligned} \tag{9.18}$$

The margin $\leq M$ could be encountered either by the change in the system capacity or the change in system load. The contributions to f_M, the frequency of encountering a margin $\leq M$, by these two respective modes of transitions are termed the generation system transitions and the load model transitions [18]

$$f_M = f_M^g + f_M^l \tag{9.19}$$

where

$$f_M^g = \sum_{i=1}^{N+1} f_v A_i \tag{9.20}$$

and

$$f_M^l = \sum_{i=1}^{N} \frac{n_i}{D} (P_v - P_{v0}) \tag{9.21}$$

It should be noted that in (9.17)–(9.21), v is a function of load level L_i selected and is so determined that $C_v \leq (L_i + M)$. The probability P_{v0} is that of capacity level corresponding to low load level such that $C_{v0} \leq (L_0 + M)$.

The CVEF and MLEF Load Models. The single exposure factor load model discussed previously assumes a random sequence of daily peaks.

This is generally not true and there is likely to be a strong sequential correlation between the daily peaks. The choice of the exposure factor is arbitrary and its nature is questionable. For example if the capacity level c_j exists for the day shown in Figure 9.3, then according to this load model a negative margin of m will exist for a duration e_i. This obviously is not an accurate representation.

Load models for accurate representation of the system load were developed in reference 18 and described in the related publications [7, 8]. In these models, the expected daily load variation curve is approximated by the mean durations at various load levels determined as a percentage of the daily peak. The expected daily curve may or may not be symmetrical. It is shown in references 7, 8, and 18 that the asymmetry does not effect the steady state probability and frequency of encountering a margin. The MLEF (Multilevel exposure factor) representation amounts to approximating the daily load variation curve as shown in Figure 9.6. For a bimodal curve this means transferring segments like d to d'. This does not alter the steady-state availability of the failure state, but the frequency is slightly decreased and the mean duration slightly increased. For example, for a given capacity level c_j, the load exceeds the available capacity twice in a day with durations e_i and d, respectively. Using this approximation, the load exceeds c_j once with duration $(e_i + d)$. Such an approximation may even give more realistic indices.

Whereas the MLEF representation assumes discrete variations in the exposure factor, the CVEF (continuously varying exposure factor) model assumes the exposure factor as a continuously varying function of the percentage of daily peak. The continuous approximation does not involve any additional difficulty and relatively little extra computational effort is required.

CAPACITY RESERVE MODEL. Four types of load models are described in references 7, 8, and 18 depending upon the manner in which low load is taken into account and the sequential interdependence of the daily peaks.

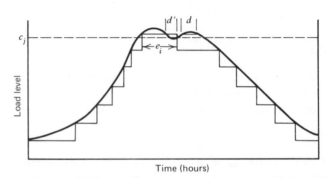

Figure 9.6 The discrete approximation of a continuously varying daily load curve.

The four possible combinations are

1. Low load assumed same on all days and the sequence of daily peaks assumed random.
2. Same as (1) above except that there is sequential interdependence between daily loads.
3. Low load different on different days and the sequence of daily loads random.
4. Same as (3) except that there is sequential interdependence between daily peaks.

Both MLEF and CVEF representations are possible for the four combinations described above. Expression for combination one, using CVEF representation is derived in this section. For other cases, the reader is referred to references 7, 8, and 18. The derivations for the availability and frequency of margin $\leqslant M$ can be understood with reference to the Figure 9.7, which shows exposure factor as a continuously varying function of the daily peak. For a given capacity level C_v, the corresponding load level L for a margin $\leqslant M$ can be determined using the relationship

$$C_v - L_v \leqslant M$$

that is,

$$L_v \geqslant C_v - M$$

The expected duration of margin $\leqslant M$ for the ith load cycle can be easily shown [18],

$$\sum_{x=0}^{n} P_{v+x}(d_{v+x} - d_{v+x-1}) \qquad (9.22)$$

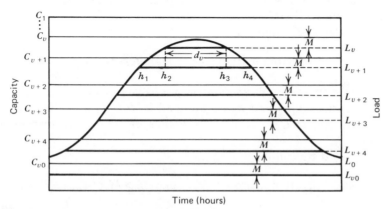

Figure 9.7 Exposure factor as a continuously varying function of the daily peak.

where d_{v+x} = the mean duration for which the load $\geq L_{v+x}$ such that
$$L_{v+x} = C_{v+x} - M$$
P_{v+x} = the probability of capacity $\leq C_{v+x}$

In (9.22), v is such that $C_v - M$ makes the first intercept with the daily load cycle and n is such that $C_{v+n} - M$ is at or just below the low load level. Assuming n_i identical peak loads L_i, the expected duration of margin $\leq M$ in the period of D days

$$= \sum_{i=1}^{N} n_i \sum_{x=0}^{n} P_{v+x}(d_{v+x} - d_{v+x-1})$$

$$= D \cdot A_M \tag{9.23}$$

where N = total number of peak loads, that is, $\displaystyle\sum_{i=1}^{N} n_i = D$

A_M = probability of margin $\leq M$

From (9.23)

$$A_M = \frac{1}{D} \sum_{i=1}^{N} n_i \sum_{x=0}^{n} P_{v+x}(d_{v+x} - d_{v+x-1}) \tag{9.24}$$

FREQUENCY OF MARGIN $\leq M$. The contribution to the frequency by the generation system transitions can be determined by finding the expected transitions out of the margin states $\leq M$. In the period $h_2 h_3$ (Figure 9.7) the load is $\geq L_v$ and therefore for the capacity level C_v, the margin $\leq M$. The generation system may transit, during this period, to a higher capacity making the margin $> M$ or to a lower capacity without change of cumulative margin state. Therefore,

The expected transitions during

$$h_2 h_3 \text{ from margin} \leq M \text{ to margin} > M = d_v f_v \tag{9.25}$$

where f_v = the frequency of encountering a capacity $\leq C_v$ or $> C_v$.

During periods $h_1 h_2$ and $h_3 h_4$, the load is $\geq L_{v+1}$ and the transitions to capacity level $> C_{v+1}$ can cause change of cumulative margin state. Therefore,

The expected transitions during

$$h_1 h_2 \text{ and } h_3 h_4 \text{ from margin} \leq M \text{ to margin} > M = (d_{v+1} - d_v)f_{v+1}$$

$$\tag{9.26}$$

Generalizing from (9.25) and (9.26), the expected transitions due to all the capacity levels, in the ith load cycle are

$$\sum_{x=0}^{n} f_{v+x}(d_{v+x} - d_{v+x-1}) \tag{9.27}$$

The expected generation system transitions out of margin $\leqslant M$ during D days are

$$\sum_{i=1}^{N} n_i \sum_{x=0}^{n} f_{v+x}(d_{v+x} - d_{v+x-1}) = f_M^g D \tag{9.28}$$

From (9.28)

$$f_M^g = \frac{1}{D} \sum_{i=1}^{N} n_i \sum_{x=0}^{n} f_{v+x}(d_{v+x} - d_{v+x-1}) \tag{9.29}$$

The load exceeds a given capacity level once a day. Therefore,

$$f_M^l = \sum_{i=1}^{N} \frac{n_i}{D} (P_v - P_{v0})$$

$$= \frac{1}{D} \sum_{i=1}^{N} n_i P_v - P_{v0} \tag{9.30}$$

From (9.29) and (9.30)

$$f_M = f_M^g + f_M^l$$

$$= \frac{1}{D} \sum_{i=1}^{N} n_i \sum_{x=0}^{n} f_{v+x}(d_{v+x} - d_{v+x-1}) + \frac{1}{D} \sum_{i=1}^{N} n_i P_v - P_{v0} \tag{9.31}$$

It should be noted that in (9.24) and (9.29) v is different for different load cycles.

EXTENSION TO THE LOSS OF ENERGY CONCEPT. The load models proposed can be readily extended to evaluate the probable curtailment of energy due to capacity shortages. Assuming the CVEF representation, the daily load variation curve is shown in Figure 9.8. The mean durations d_j are stored in the computer in discrete steps and the mean duration between any two discrete steps can be computed by linear interpolation. The magnitude of load corresponding to the mean duration d_j is called the subload level L_{ij} where i refers to peak load level.

For a capacity level C_v, the energy curtailed is equal to the area bounded by the expected load variation curve and the mean duration d_v

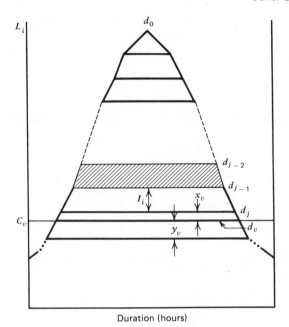

Figure 9.8   Expected daily load variation curve for the *i*th peak.

for which the load exceeds C_v. The area of the shaded portion is

$$= \frac{1}{2}(d_{j-1}+d_{j-2})I_i$$

where I_i = the interval in MW between the successive subload levels, for the *i*th peak load
   $= L_i \cdot p/100$

where L_i = the *i*th peak load, MW
   p = the interval between successive subload levels as a percentage of the daily peak

The energy curtailed

$$= \frac{1}{2}(d_j+d_v)x_v + \frac{1}{2}(d_j+d_{j-1})I_i + \cdots + \frac{1}{2}(d_0+d_i)I_i$$

$$= \frac{1}{2}\left[(d_v+d_j)(I_i-y_v)+(d_j+d_{j-1})I_i + \cdots + (d_0+d_1)I_i\right]$$

$$= \frac{1}{2}(d_v \cdot x_v - d_j \cdot y_v) + I_i \cdot E_j$$

where

$$E_j = \sum_{k=0}^{j} d_k$$

= the cumulative total of the subload mean durations up to jth sublevel such that $L_{ij} \geqslant C_v$

where L_{ij} = the value of the jth sublevel of the ith peak, MW

The expected curtailment of energy, given C_v, L_i

$$= P_v' \cdot \left[\frac{1}{2} (d_v \cdot x_v - d_j \cdot y_v) + I_i \cdot E_j \right]$$

The total expected curtailment of energy in a period of D days

$$EN = \sum_{i=1}^{N} n_i \sum_{v=1}^{R} P_v' \left[\frac{1}{2} (d_v \cdot x_v - d_j \cdot y_v) + I_i \cdot E_j \right] \qquad (9.32)$$

where P_v' = the exact state availability of capacity level C_v
n_i = number of occurrences of peak L_i, $1 = 1, 2 \ldots N$

such that

$$D = \sum_{i=1}^{N} n_i$$

and $L_{ij} \geqslant C_v$.

The expression (9.32) is suitable for digital computation and its execution is very fast. The value of EN given by this expression may be multiplied by a suitable cost factor in \$/MW-Hr to get the expected loss in dollars. To act as an index of reliability EN must be normalized by dividing it by the total energy required by the system. This quantity is given by the expression

$$ENP = \sum_{i=1}^{N} n_i \left[I_i \left(\frac{1}{2} d_n + E_{n-1} \right) + 24 (L_i - n \cdot I_i) \right] \qquad (9.33)$$

where n = the sublevel just at the low load. This quantity, which is rather small, may be subtracted from unity to get what is conventionally called the Energy Index of Reliability. Thus

$$EIR = 1.0 - EN/ENP \qquad (9.34)$$

SEQUENTIAL CORRELATION OF DAILY PEAKS. A number of models have been proposed and analyzed in references 7 and 18. The essential difference between the various load models is the assumptions regarding low load and the sequential correlation of daily peaks. The analysis leads to an interesting conclusion that, if the low load can be considered of constant magnitude, the numerical values of the probability and frequency of margin states are the same whether the sequence of peak is assumed random or correlated. For most systems, the probability of having capacity as low as the low load period is very small and therefore the error in the numerical indices of reliability because of the assumption of constant low load is insignificant. The model discussed here is thus adequate in most situations of interest. If, however, low load cannot be assumed constant, the numerical indices are effected and appropriate load models [7, 18] can be employed.

9.3 ASSESSMENT OF OPERATING RESERVE

The operating reserve evaluation is concerned with the ability of the generation system to meet the load within the next few hours. If a generating unit fails, additional capacity can be brought in after a time equal to the start up time of the reserve units. This time known as lead time, delay time or start up time is different for different types of units. It is of the order of a few minutes for hydraulic, gas turbines; for the thermal units on cold standby it may be 4–24 hours. One way of reducing this time is to keep the boilers banked; these units are called hot reserve units. The reserve connected to bus ready to take load is called spinning reserve. This spinning reserve together with the rapid start and hot reserve units is called operating reserve. The basic problem is to decide how much reserve to have so that the load can be satisfied with reasonable level of risk. Three methods have been proposed for the assessment of the operating reserve and an excellent review of these is provided in reference 15. These methods are briefly discussed in this section.

9.3.1 Basic PJM Method

This method was first described in 1963 by a group associated with the Pennsylvania–New Jersey–Maryland interconnection [2]. The index computed is the probability of having insufficient capacity in operation at a future time equal to the time needed to bring in additional generating capacity. It is assumed that there is enough installed capacity and it is just a matter of time before the additional capacity can be brought in to share the load. The present state of the system is assumed known and the start up time of all the stand-by units is considered the same. The procedure for computation is basically similar to the static reserve evaluation. The essential difference is the time.

Generation System Model. The concepts will be illustrated using a two state unit. There is, however, no additional difficulty in including units with derated states. The probability of a two-state unit being down at time T, given that it is operating at $t=0$ is [23]

$$p_d(T) = \frac{\lambda}{\lambda+\mu}\left[1-e^{-(\lambda+\mu)T}\right] \tag{9.35}$$

where λ, μ are the failure and repair rates of the unit.

If $(\lambda+\mu)T \ll 1$, then (9.35) can be approximated as

$$p_d(T) \simeq \frac{\lambda}{\lambda+\mu}\cdot(\lambda+\mu)T$$

$$= \lambda T \tag{9.36}$$

The expression (9.36) can also be obtained by assuming that there is no repair in $(0, T)$ and that T is small. If T is the start up time of additional capacity, then λT is the probability of loosing capacity and not being able to replace it and is called ORR, that is, the outage replacement rate. After computing the ORRs for the units scheduled at time $t=0$, the probability of having various levels of capacity outage at T can be calculated either by the product rule or more generally by the algorithm described for the static reserve evaluation.

Load Model and Risk Calculation. The load model for the operating reserve evaluation is the forecast load at T. The risk, that is, the probability of having insufficient capacity at T, is

$$R = \sum_i Pr(\text{load at } T = L_i)\, Pr(\text{capacity at } T < L_i) \tag{9.37}$$

where L_i is the value of forecast load. The continuous distribution of forecast values is approximated by a discrete distribution. If no uncertainty in forecast load is assumed, then there is only one value of L_i and

$$R = Pr(\text{capacity at } T < \text{load at } T) \tag{9.38}$$

The computed R is then compared with a maximum tolerable risk R_{ref} to decide whether the system is adequately secure. If the computed risk is more or less than the reference value, suitable action can then be taken. At present the selection of R_{ref} is not simple and is generally based on judgment and past experience.

9.3.2 Modified PJM Method

The basic PJM method can be modified to incorporate the effect of rapid start and hot reserve units [10]. The models for the rapid start and hot

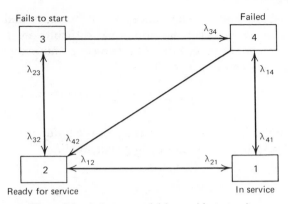

Figure 9.9 A 4-state model for rapid start units.

reserve units are shown in Figures 9.9 and 9.10. The state transition diagrams are self-explanatory and the transition rates are given by

$$\lambda_{ij} = \frac{n_{ij}}{T_i} \tag{9.39}$$

where λ_{ij} = transition rate from state i to state j
n_{ij} = number of transitions from state i to state j during the time T_i
spent in state i

   Assuming the times in the various states to be exponentially distributed, the state differential equations can be written as

$$\dot{P}_i(t) = \sum_{j \neq i} P_j(t) \lambda_{ji} - P_i(t) \sum_{j \neq i} \lambda_{ij} \tag{9.40}$$

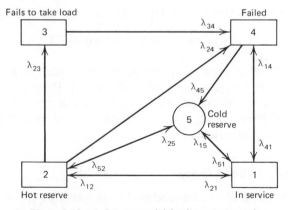

Figure 9.10 A 5-state model for hot reserve units.

The set of differential equations (9.40) can be solved using the initial conditions to calculate the probabilities of being in various states. Denoting the start up times of rapid start, hot reserve, and cold reserve units by t_r, t_h, and t_c respectively, the modified PJM method proceeds in the following steps:

1. Using the scheduled units, that is, the units on line at $t=0$, the probability of having insufficient generation at t_r is computed using the basic PJM approach, that is,

$$R(t_r)=\text{risk during the interval } (0, t_r)$$

$$=\text{probability of having insufficient generation at } t_r$$

2. During $(0, t_r)$ the rapid start units are assumed to be in ready for service with probability of unity. Using the differential equations (9.40) and initial condition, $P_2(t_r^+)=1\cdot0$, the probabilities of finding the rapid start units in states 3 and 4 at time t_h are determined. Denoting these probabilities by P_{23} and P_{24}, the probability of having the rapid start unit failed at t_h is $(P_{23}+P_{24})$. These probabilities of the rapid start units are combined with the probabilities of scheduled units at t_h to develop the generation model at t_h. The generation model at t_h consists of the units in operation at $t=0$ and the rapid start units that are assumed to become available at $t=t_r^+$. This generation model is combined with the load at t_h to find the probability

$$R(t_h)=\text{probability of having insufficient generation at } t_h$$

Also $R(t_r^+)$ is computed by considering all the units in operation at $t=0$ and using their probabilities along with the probabilities of the rapid start units at $t=t_r^+$, which is, in fact, zero time for these units.

3. At time t_h^+, the hot reserve units are assumed to become available. The probability of insufficient capacity at t_h^+ is calculated by modifying the generation model at t_h by combining with the probabilities of hot reserve units at t_h^+ which is, in effect, zero time for these units since in the interval $(0, t_h)$, the hot reserve units are assumed in the hot reserve state with a probability of unity. The probabilities of hot reserve units at t_c are computed with $P_2(t_h^+)=1.0$ and using differential equations or matrix multiplication technique [23]. The generation system at t_c, therefore, consists of (a) units in operation at $t=0$, having operated for t_c, (b) rapid start units having operated for (t_c-t_r) and (c) the hot reserve units having operated for (t_c-t_h). This generation model is then combined with the load at t_c to find the probability of insufficient capacity at t_c, that is, $R(t_c)$. The risk at t_c according to reference 10 is

given by

$$R = R(t_r) + [R(t_h) - R(t_r^+)] + [R(t_c) - R(t_h^+)] \qquad (9.41)$$

The risk given by (9.41) is used as an index in the modified PJM method. It appears, however, difficult to assign a physical significance to this index. Also it appears that reference 10 does not incorporate the models for rapid start and hot reserve units in a realistic manner. Denoting $P_{ij}(t)$ as the probability of being in state j starting in state i, reference 10 appears to compute the probability of a rapid start unit as

$$P_d(t) = P_{23}(t) + P_{24}(t) \qquad (9.42)$$

and

$$P_u(t) = 1 - P_d(t) \qquad (9.43)$$

where $P_u(t)$, $P_d(t)$ are the probabilities of the rapid start units being up and down, respectively, at time t. The initial condition assumed is $P_2(0) = 1.0$, which does not reflect the fact that at $t = 0$, the unit was commanded to start. Reference 1 suggests an improvement in the procedure for incorporating the effect of rapid start and hot reserve units. When a rapid start unit, for example is commanded to start, it either starts (state 1) or fails to do so (state 3). Denoting the probability of starting or failing to start by s and f, respectively,

$$s = \frac{\lambda_{21}}{\lambda_{21} + \lambda_{23}} \qquad (9.44)$$

$$f = \frac{\lambda_{23}}{\lambda_{21} + \lambda_{23}} \qquad (9.45)$$

The probabilities of various states are now computed with the probabilities of starting in states 2 and 3 as s and f, respectively. After computing $P_{ij}(t)$, the probability of the unit being down and up given the period of need are computed as

$$P_{\text{out}} = \frac{P_{23}(t) + P_{24}(t)}{P_N} \qquad (9.46)$$

and

$$P_{\text{in}} = \frac{P_{21}(t)}{P_N} \qquad (9.47)$$

where

$$P_N = P_{21}(t) + P_{23}(t) + P_{24}(t)$$

The probabilities calculated by (9.46) and (9.47) are then used in the risk computation instead of $P_d(t)$ and $P_u(t)$ given by (9.42) and (9.43).

9.3.3 Security Function Method

The security function method was proposed in reference 14 and later modified and expanded in several publications [15]. Basically this method calculates the probability of system trouble as a function of time. The time span of computation is the lead time required for the modification of the system operating configuration to achieve improved system security. The form of security function suggested in reference 14 is

$$S(t) = \sum_i P_i(t) W_i(t) \qquad (9.48)$$

where $P_i(t)$ = probability of the system being in state i at time t
$W_i(t)$ = probability that the system configuration of state i results in system trouble

Equation 9.48 in its general form can be applied to the entire set of components comprising a bulk power system. When applied to the operating reserve problem $S(t)$ indicates the probability of insufficient capacity at time t into future. The function $S(t)$ is examined for a time period equal to the lead time, that is, the time to start and synchronize additional capacity. If the security function is exceeding a predefined reference value, then a decision to start additional capacity can be taken. Likewise if the system appears too secure, appropriate generating capacity may be taken out for economic operation. This method treats the standby generators in a rational manner and in conformity with the normal operating practices. In the modified PJM method, the standby generators are started only when a scheduled unit fails. The amount of standby generators are shut down when the scheduled unit has been repaired. In contrast to the modified PJM method, there is no difficulty in interpreting the security function $S(t)$. It should be noted that when only the generating system is considered and when t is the shortest start up time for stand-by generators then the risk obtained by the security function method is the same as the basic PJM method.

9.3.4 Frequency and Fractional Duration Method

The frequency and duration method for short-term reliability calculation originated in reference 18 and is also described in reference 22. The previously described methods calculate the pointwise probability of capacity deficiency. Although the security function method calculates $S(t)$ over the entire period, the total interval is not considered at a time. The frequency and duration method in addition to the pointwise probability of generation deficiency, also calculates two additional interval related indices, interval frequency, and fractional duration.

Basic Concepts. The entire sample space X can be partitioned into disjoint subsets X^+ and X^-. Whenever the system enters any state contained in X^+, this subset of states is said to have been encountered. The following indices can now be defined.

TIME SPECIFIC PROBABILITY OF X^+. This is the probability of the system being in any state contained in X^+ at time t,

$$P_+(t) = \sum_{i \in X^+} P_i(t) \tag{9.49}$$

where $P_i(t)$ is the probability of being in state i at time t. When X^+ is constituted by states indicating system trouble, (9.49) becomes identical with (9.48).

FRACTIONAL DURATION. The fractional duration of X^+ in the interval (t_1, t_2) is defined as the expected proportion of (t_1, t_2) spent in X^+. Denoting fractional duration by $D_+(t_1, t_2)$,

$$D_+(t_1, t_2) = \frac{1}{t_2 - t_1} \sum_{i \in X^+} \int_{t_1}^{t_2} P_i(t) \, dt$$

$$= \frac{\int_{t_1}^{t_2} P_+(t)}{t_2 - t_1} \tag{9.50}$$

INTERVAL FREQUENCY. The interval frequency $F_+(t_1, t_2)$ is defined as the expected number of encounters of X^+ in (t_1, t_2).

$$F_+(t_1, t_2) = \sum_{i \in X^-} \int_{t_1}^{t_2} P_i(t) \sum_{j \in X^+} \lambda_{ij} \, dt \tag{9.51}$$

where λ_{ij} is the constant transition rate from state i to state j.

Application to Operating Reserve. The relationships (9.49)–(9.51) are general and can be applied to the entire system or parts thereof. The application of these concepts to operating reserve evaluation involves the following two basic steps.

GENERATION SYSTEM MODEL. The generation system model depicts the time-specific probability, the fractional duration and the interval frequency as functions of the cumulative capacity outages, that is, capacity outages equal to or greater than specific values. The numerical techniques for developing generation system model are described in references 18 and 22.

GENERATION RESERVE MODEL. The load is assumed to be forecast with probability one and to stay constant over the hourly intervals. If, however, the load is forecast with a certain probability distribution, there is no additional difficulty in incorporating this and also if a closer representation is required, the intervals over which the load is assumed constant can be made as small as desired. Since the load is assumed to exist at a certain number of discrete levels and as the capacity states are also discrete, the operating or generation reserve which is capacity minus the load would also exist in discrete levels. This can be illustrated by assuming the load for four hours as

Hour	0–1	1–2	2–3	3–4
Load	20	40	50	60

The hours will now be indicated by interval numbers, for example, 0–1 will be denoted by interval # 1. This forecast load is combined with the generation model, the resulting generation reserve will be as shown in Table 9.1. The boundary of any cumulative margin, that is, a margin equal to or less than a specified value can now be drawn. The boundary of capacity deficient states, for example, is shown in Table 9.1. If the interval frequency of encountering capacity deficiency is to be determined, then $F_+(0,4)$ will be determined where X^+ will contain all the states below the thick line.

In general, denoting the capacity associated with the ith cumulative capacity outage state by C_i, the boundary for cumulative reserve margin M, that is, a margin equal to or less than M MW, corresponding to the load during the jth interval L_j is fixed by the relationship

$$C_i - L_j \leqslant M$$

That is,

$$C_i \leqslant L_j + M \qquad (9.52)$$

Table 9.1 *The Generation Reserve Model of the Example*

	Interval #			
	1	2	3	4
Load	20	40	50	60
Cumulative Capacity Outage				
0	55	35	25	15
25	30	10	0	−10
50	5	−15	−25	−35
75	−20	−40	−50	−60

The expressions for the different indices can be written using the following notation:

$T=$ the lead time

$P_i(t)=$ the probability of the ith cumulative capacity outage state at time t

$D_i(t_1, t_2)$,
$F_i(t_1, t_2)=$ the fractional duration and the interval frequency of encountering the ith cumulative capacity outage state in the interval (t_1, t_2)

$P_M(T)=$ the probability of the cumulative margin M at the end of lead time

$D_M(0, T)$,
$F_M(0, T)=$ the fractional duration and the interval frequency of encountering the cumulative margin M in the interval $(0, T)$, that is, during the lead time.

Probability. The expression for time specific probability of the cumulative margin M is straightforward:

$$P_M(T)=P_i(T)\tag{9.53}$$

such that

$$C_i \leqslant L(T)+M$$

where $L(T)=$ the load at time T.

The Fractional Duration. Denoting the time at the end of jth interval by t_j,

$$D_M(0, T)=\sum_{j=1}^{m} D_i(t_{j-1}, t_j)\tag{9.54}$$

where m is the total number of intervals in the lead time T, and the cumulative capacity outage state i during the interval j is determined using (9.52).

The Interval Frequency. The state of margin equal to or less than M may be encountered either due to a decrease in capacity or increase in load. There are, therefore, two components of interval frequency, the generation system transition $F_M^g(0, T)$ and the load transitions, $F_M^l(0, T)$ such that

$$F_M(0, T)=F_M^g(0, T)+F_M^l(0, T)$$

Now

$$F_M^g(0, T) = \sum_{j=1}^{m} F_i(t_{j-1}, t_j)$$

and

$$F_M^l(0, T) = \sum_{j=1}^{m} \left[P_k(t_j) - P_i(t_j) \right]$$

such that $C_i \leqslant M + L_j$ at the end of the jth interval and

$$C_k \leqslant M + L_j' \text{ at the beginning of the } (j+1)\text{th interval.}$$

Finally,

$$F_M(0, T) = \sum_{j=1}^{m} F_i(t_{j-1}, t_j) + \sum_{j=1}^{m} \beta \left[P_k(t_j) - P_i(t_j) \right] \qquad (9.55)$$

where $\beta = 1$ if $[P_k(t_j) - P_i(t_j)]$ is positive
$\quad\quad\ = 0$ otherwise

Expressions 9.53–9.55 can be used to determine the three indices for cumulative margin M. If M is such that it defines the capacity deficiency states the three indices are the three risk indices.

9.4 INTERCONNECTED SYSTEMS

The interconnection of a power system to one or more power systems generally improves the generating capacity reliability. When the power system suffers a loss of load, assistance is generally available from the systems to which it is interconnected. The benefits of interconnection result from the diversity in the occurrence of the peak loads and the outages of capacity in different systems.

In evaluating the reliability of interconnected systems, the load and generation in each system are assumed to be connected to a common bus and the tie lines are assumed to connect these buses together. This is shown schematically in Figure 9.11. This means that within each system, the transmission system is assumed capable of transferring the available generation to the points of demand. Also when needed generation is made available to a system from a neighboring system, it is assumed that the intratransmission system can then properly distribute this capacity.

The type of agreement with the assisting systems effects the evaluation of the reliability of an interconnected system. This discussion assumes a

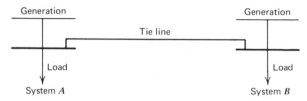

Figure 9.11 Schematic diagram of system *A* connected to system *B*.

basic agreement that one system helps the other as much as it can without curtailing its own load. The concept may, however, be easily extended to cover other agreements. The discussion in this section is limited to a system *A* connected to system *B*, which is generally called a two area problem. For a detailed discussion of this problem and that on multi-area problems, the reader is referred to references 17, 18, and 3–5. The methods of reliability analysis discussed in this section were developed in references 17 and 18 and later described in references 3–5.

9.4.1 *Independent Load Models*

The relationships for the probability and frequency of negative margins (load loss) in system *A* connected to system *B* are developed assuming the generation and load models in the two systems to be stochastically independent. Assuming the capacity and load in each system to exist at discrete levels, the margin state, which is capacity available less the load on the system, would also exist at discrete levels in each system. In Figure 9.12, M_a, M_b contain the margin states in systems *A* and *B*, respectively,

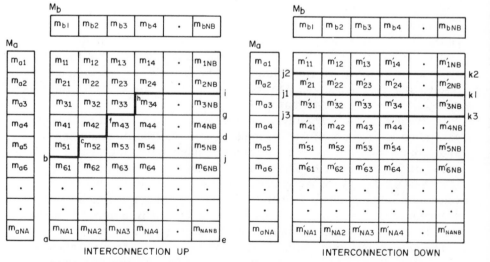

Figure 9.12 Effective margin states matrices for system *A* connected to system *B*.

without including the effect of interconnection assistance. These states are arranged in the order of decreasing reserve, that is,

$$m_{a1} > m_{a2} > \cdots > m_{ai} > \cdots > m_{aNa}$$

and

$$m_{b1} > m_{b2} > \cdots > m_{bj} > \cdots > m_{bNB}$$

The effective margin states in system A, that is, when the tie line is in operation, are given by the elements of matrix M,

$$m_{ij} = m_{ai} + h_{ij} \qquad (9.56)$$

where h_{ij} is either the help available to system A from system B or it is the help required by system B from system A. In the latter case h_{ij} has a negative sign. If no assistance is possible from one system to the other, $h_{ij} = 0$. The maximum of h_{ij} is limited by the tie line capability. The effective margin states in system A while the tie line is out are given by the elements of M',

$$m'_{ij} = m_{ai} \qquad (9.57)$$

Equations 9.56 and 9.57 define the boundaries in M and M', respectively, of any effective cumulative margin. In this discussion, m with proper subscript represents an exact margin and M with the same subscript denotes the corresponding cumulative margin, for example, M_{ij} means a margin equal to or less than m_{ij}. The probability and frequency equations for the negative cumulative margin are derived and for the equations for any margin, positive or negative, the reader is referred to reference 3.

Probability of Margin Equal or Less Than N. The probability of a margin equal to or less than N is simply the sum of the probabilities of the margins states comprising this cumulative state. The equation for this probability is easily seen to be

$$P_N = \sum_{l,\,k} \left[P_{a(l)} - P_{a(l+1)} \right) \left(A_{ab} P_{b(k)} + \bar{A}_{ab} \right] \qquad (9.58)$$

where $P_{a(l)},\, P_{b(k)}$ = the probabilities of cumulative margins $M_{a(l)}$ and $M_{b(k)}$, respectively

A_{ab} = the availability of the tie line between A and B

$\bar{A}_{ab} = 1 - A_{ab}$

l, k = the indices defining the boundary of the margins less than or equal to N. The index l indicates the margin in array M_a and k indicates the corresponding margin in M_b to give an effective margin $\leqslant N$. For example the boundary in Figure 9.12 is identified by $(NA, 1)$, $(NA - 1, 1), \ldots, (6, 1),\ (5, 2),\ (4, 3),$ and $(3, 4)$.

Equation (9.58) can be easily seen to be an application of conditional probability theorem [23].

Frequency.   Define

$f_{a(l)}$, $f_{b(k)}$ = the frequencies of encountering the cumulative margin states $M_{a(l)}$ and $M_{b(k)}$, respectively

λ_{ab}, μ_{ab} = the mean failure and repair rates of the tie line

and

f_N = the frequency of encountering an effective margin in system A, equal or less than N

System A can transit from one effective margin state to another in any of the following ways.

1. *Capacity or load transitions in System A.* System A will shift vertically in M when the interconnection is in operation and in M' when the interconnection is on forced outage.
2. *Capacity or load transitions in System B.* Due to transitions in system B, system A will transfer horizontally from one effective state to another in the matrix M, when the interconnection is up. With the interconnection in the down state, the system A will transfer horizontally in the matrix M'. These latter transitions do not ultimately reflect into the effective operation of system A.
3. *Failure or repair of the interconnection.* When the interconnection fails or is repaired, system A will transit from a state in M' to the corresponding state in M and vice versa.

The frequency of encountering a cumulative margin in system A equals the expected transitions per unit time across the boundary defining that cumulative margin plus the transitions per unit time associated with the boundary states. The boundary states are a subset of the cumulative margin when the interconnection is down but leave this set as a result of repair of the interconnection.

Summing the frequency due to the three modes of transition gives the following relationship.

$$f_N = \sum_{l,k} \left\{ \left[f_{a(l)} - f_{a(l+1)} \right] \left[A_{ab} P_{b(k)} + \overline{A}_{ab} \right] \right.$$

$$\left. + \left[P_{a(l)} - P_{a(l+1)} \right] \left(f_{b(k)} + \left[1 - P_{b(k)} \right] \lambda_{ab} \right) A_{ab} \right\} \qquad (9.59)$$

Equation 9.59 can be easily derived by the application of the conditional frequency formula [23],

$$f_N = Fr(N/TL \text{ Up})A_{ab} + Fr(N/TL \text{ Down})A_{ab}$$

$$+ \left[P(N/TL \text{ Down}) - P(N/TL \text{ Up}) \right] A_{ab}\lambda_{ab} \qquad (9.60)$$

where

$$Fr(\cdot), P(\cdot) \text{ are the frequency and probability of } (\cdot).$$

$$TL = \text{Tie line}$$

Now

$$Fr(N/TL \text{ up}) = \sum_{l,k} \left\{ \left[f_{a(l)} - f_{a(l+1)} \right] P_{b(k)} + \left[P_{a(l)} - P_{a(l+1)} \right] f_{b(k)} \right\}$$

$$(9.61)$$

$$Fr(N/TL \text{ down}) = \sum_{l,k} \left[f_{a(l)} - f_{a(l+1)} \right] \qquad (9.62)$$

and

$$P(N/TL \text{ Down}) - P(N/TL \text{ Up}) = \sum_{l,k} \left(1 - P_{b(k)}\right) \qquad (9.63)$$

It can be easily seen that substitution of (9.61)–(9.63) into (9.60) will yield (9.59).

9.4.2 Correlated Load Models

In the discussion in Section 9.4.1., the load models in the two systems are assumed statistically independent. It is, however, more likely that the loads in the two interconnected systems will bear a correlation. This section develops the relationships for the probability and frequency of a cumulative margin, assuming the loads in the two systems to be perfectly correlated.

Let

$(L_{ax}, L_{bx}) =$ perfectly correlated load levels in systems A and B where $x = 1, 2, \ldots, n$

$M_a^x, M_b^x =$ the reserve margin state arrays of systems A and B corresponding to the load condition (L_{ax}, L_{bx})

The margin state arrays M_a^x and M_b^x can be obtained by subtracting loads L_{ax} and L_{bx} from the capacity states of the systems A and B, respectively. The arrays M_a^x and M_b^x are shown in Figure 9.13 where the margin states are arranged in the decreasing order of magnitude.

LOAD CONDITION (L_{ax}, L_{bx})

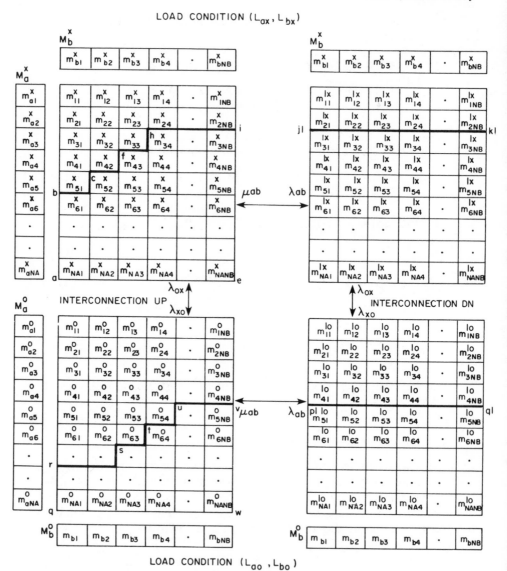

Figure 9.13   Effective margin states matrices for system A connected to system B.

As before c and m^x with proper subscripts represent exact capacity and conditional exact margin state and C and M^x with the same subscript denote the corresponding cumulative states. For example, C_{ai} means a capacity equal to or less than c_{ai}, and similarly M_{ij}^x means a margin equal to or less than m_{ij}^x. The effective margin states in system A given the load condition (L_{ax}, L_{bx}) and the interconnection in the up state are given by the elements of matrix M^x of Figure 9.13,

$$m^x_{ij} = m^x_{ai} + h^x_{ij} \qquad (9.64)$$

where h^x_{ij} is either the help available to system A from system B or it is the help required by system B from system A, given the load condition (L_{ax}, L_{bx}).

The effective margin states in system A given the load condition (L_{ax}, L_{bx}) and with the interconnection down are given by the elements of matrix M^{1x}

$$m^{1x}_{ij} = m^x_{ai} \qquad (9.65)$$

Equations 9.64 and 9.65 define the boundaries of effective cumulative margin states in M^x and M^{1x}, respectively. The effective margin states in system A, given the low load condition (L_{ao}, L_{bo}) and the interconnection in the up state are given by the matrix M^o (see Figure 9.13), and the effective margin states for the above condition with the interconnection in the down state are given by M^{1o}.

Probability of Margin Equal or Less Than N.   Using (9.58)

$$P_{(N/x)} = \sum_{lx,\,kx} \left[P_{a(l)} - P_{a(l+1)} \right] \left[A_{ab} P_{b(k)} + \bar{A}_{ab} \right] \qquad (9.66)$$

where $P_{(N/x)}$ = probability of an effective margin $\leqslant N$, given the load condition (L_{ax}, L_{bx})

lx, kx = the indices defining the boundary of N in M^x, for example, if the boundary is "$abcfhie$", the indices are $(NA, 1) - (6, 1)$, $(5, 2)$, $(4, 3)$, and $(3, 4)$ (see Figure 9.13)

$P_{a(l)}, P_{b(k)}$ = probabilities of M^x_{al} and M^x_{bk}, given the load condition (L_{ax}, L_{bx})

Since

$$m^x_{al} = c_{al} - L_{ax}$$

and

$$m^x_{bk} = c_{bk} - L_{bx}$$

it can be seen that $P_{a(l)}, P_{b(k)}$ are probabilities of C_{al} and C_{bk}, respectively.

Using conditional probability, for n load levels and low load level $(x = 0)$,

$$A_N = \sum_{x=0}^{n} A_x P_{N/x} \qquad (9.67)$$

where A_x = probability of load condition (L_{ax}, L_{bx}).

Frequency. The two different modes of state transition in system A for a given load condition (L_{ax}, L_{bx}) are

1. The generation system transitions in system A or system B.
2. The state transition due to the failure or the repair of the tie line.

The frequency of encountering any cumulative effective margin in system A due to these two modes can be determined using (9.59),

$$f_{(N/x)} = \sum_{lx, kx} \left\{ \left(f_{a(l)} - f_{a(l+1)} \right) \left[A_{ab} \cdot P_{b(k)} + \bar{A}_{ab} \right] \right.$$

$$\left. + \left[P_{a(l)} - P_{a(l+1)} \right] \left[f_{b(k)} + \left[1 - P_{b(k)} \right] \lambda_{ab} \right] A_{ab} \right\} \qquad (9.68)$$

where $f_{(N/x)} =$ the frequency of encountering an effective margin $\leqslant N$, given the load condition (L_{ax}, L_{bx})
$f_{a(l)}, f_{b(k)} =$ the frequencies of encountering M_{al}^x and M_{bk}^x, respectively, given the load condition (L_{ax}, L_{bx})
$=$ the frequencies of encountering C_{al} and C_{bk}, respectively

The frequency of encountering the effective margin $\leqslant N$ with the load condition (L_{ax}, L_{bx}) is

$$f_N^x = A_x \cdot f_{(N/x)}$$

The frequency due to the two modes listed previously is given by the summation over all the load conditions, that is,

$$f_{CN} = \sum_{x=1}^{n} f_N^x + f_N^o$$

This f_{CN} represents the contribution to f_N (the frequency of encountering the margin $\leqslant N$) due to the generation transitions and also takes into account transitions due to failure or repair of the interconnection. It does not, however, take into account the contribution due to the load transitions.

THE CONTRIBUTION DUE TO THE LOAD TRANSITIONS. The peak loads are assumed to be followed by the low load period. Thus the system can transit from a load condition (L_{ai}, L_{bi}) to (L_{ao}, L_{bo}) and again to some load condition (L_{aj}, L_{bj}). It should be kept in mind that there are no interpeak transitions, that is, the system cannot transit directly from (L_{ai}, L_{bi}) to (L_{aj}, L_{bj}). The contribution to f_N will thus result from the transition of system A from a given load condition to the low load condition and vice versa.

Let the boundary of N be represented by "$a\,b\,c\,f\,h\,i\,e$" and "$j\,l\,k\,l$" in M^x and M^{1x}, respectively. The corresponding boundary in M^o and M^{1o} may be represented by "$q\,r\,s\,t\,u\,v\,w$" and "$p\,l\,q\,l$". The boundary states as a result of the load transitions are those states included in the boundaries "$a\,b\,c\,f\,h\,i\,e$" and "$j\,l\,k\,l$" but not in "$q\,r\,s\,t\,u\,v\,w$" and "$p\,l\,q\,l$". Let these states be represented by a set S. The contribution due to the transition from load condition (L_{ax}, L_{bx}) to the low load condition (L_{ao}, L_{bo}) are given by

$$f_x = \sum_{s \in S} A_{(s/x)} \cdot A_x \cdot \lambda_{xo}$$

where $A_{(s/x)}$ = the probability of a boundary state $s \in S$, given the load condition (L_{ax}, L_{bx})

λ_{xo} = the transition rate from the load condition (L_{ax}, L_{bx}) to (L_{ao}, L_{bo})

It can be seen that

$$\sum_{s \in S} A_{(s/x)} = A_{(N/x)} - A_{(N/o)}$$

Therefore,

$$f_x = \left[A_{(N/x)} - A_{(N/o)} \right] \cdot A_x \cdot \lambda_{xo}$$

The total contribution can be found by summation over all the load conditions, and is

$$\sum_{x=1}^{n} f_x$$

Adding this to f_{CN}

$$f_N = \sum_{x=1}^{n} \left\{ f_{(N/x)} \cdot A_x + \left[A_{(N/x)} - A_{(N/o)} \right] \cdot A_x \cdot \lambda_{xo} \right\} + f_N^o$$

$$= \sum_{x=1}^{n} A_x \cdot \left\{ f_{(N/x)} + \left[A_{(N/x)} - A_{(N/o)} \right] \lambda_{xo} \right\} + f_N^o \qquad (9.69)$$

The step by step procedure for evaluating the reliability indices in system A, with correlated load models is outlined as follows.

1. The low load condition (L_{ao}, L_{bo}) is selected first and from each capacity state in C_a, L_{ao} is subtracted to obtain M_a^o. M_b^o is obtained in a similar manner. The boundary of the failure state in M^o and M^{1o} can be fixed using (9.64) and (9.65).

2. $A_{(N/o)}$ and $f_{(N/o)}$ are evaluated using equations (9.67) and (9.68). Here N represents the effective margin $\leqslant$ the first negative margin in M^o or M^{1o}. If the low loads in the two systems are assumed zero,

$$A_{(N/o)}=0$$

$$f_{(N/o)}=0$$

3. $A_{(n/x)}$ and $f_{(N/x)}$ for every load condition are evaluated in the same manner as outlined for the low load state.

4. The probability and the frequency of the failure state in system A can be finally found using the (9.67) and (9.69). Assuming the low load level in both the systems to be zero, these equations simplify to

$$A_- = \sum_{x=1}^{n} A_x \cdot A_{(N/x)}$$

and

$$f_- = \sum_{x=1}^{n} A_x \left[f_{(N/x)} + A_{(N/x)}/e \right]$$

where e = the load exposure factor.

9.4.3   System Studies

The techniques described in Sections 9.4.1. and 9.4.2. have been implemented in a computer program [21] and several studies based on this program have been reported in references 17, 18, and 3–5. A typical study, the effect of tie line capacity on risk level in system A, is reported here.

A system designated A is assumed to be connected to an identical system B by a single tie line. The mean failure and repair rates of the tie line are assumed to be 0.01 and 2.5 per day, respectively. The description of the generation system and load model in each system is provided in Table 9.2.

Table 9.2   *Generation System*

No. of Identical Units	Unit Size (MW)	Mean Down Time (Years)	Mean Up Time (Years)
1	250	0.06	2.94
3	150		
2	100		
4	75		
9	50		
3	25		

Total number of units = 22
Total installed capacity = 1725 MW

Load System
Exposure factor = 0.5 day
Period = 20 days

Load Condition

(x)	(MW, MW)	No. of Occurrences
1	(1450, 1450)	8
2	(1255, 1255)	4
3	(1155, 1155)	4
4	(1080, 1080)	4

The low load level in both the systems was assumed to be at zero MW.

The study was carried out by varying the tie capability from 25 to 625 MW. The mean failure and repair rates of the tie were maintained at 0.01 and 2.5 per day. The results of this study are shown in Figures 9.14 and 9.15. The curves representing the correlated load models are labeled 1 and those corresponding to the independent load models by 2. With the lower

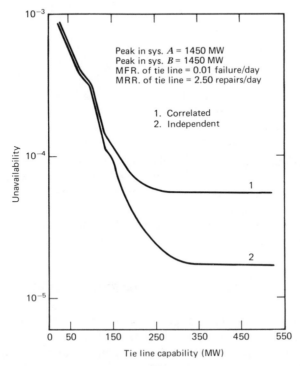

Figure 9.14 Variation of risk level (unavailability) in system *A* with the variation of tie line capability.

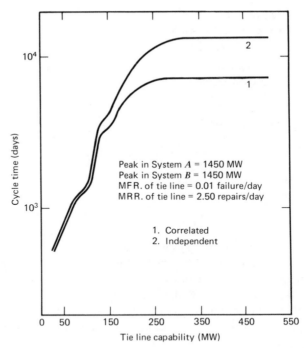

Figure 9.15 Variation of risk level (cycle time) in system *A* with the variation of tie line capacity.

values of tie line capability, the system is closer to being isolated rather than interconnected. The effect of the interconnection is, therefore, not significant and the difference between the two sets of results is not discernible. As the tie capacity is increased, the interconnection becomes more effective, and, around 125 MW, the two results begin to deviate significantly. It can be seen from Figures 9.14 and 9.15 that beyond 250 MW there is no marked improvement in reliability indices for curve 1. This is then the practical limit for tie capability with the correlated load models and in this case it is reasonably close to the independent load models condition. The limiting values of unavailability and cycle time for the two cases are, however, significantly different. The independence assumption gives optimistic results as compared with the correlated load models.

9.5 TRANSMISSION AND DISTRIBUTION SYSTEMS

A number of techniques have been proposed for the quantitative evaluation of transmission and distribution system reliability. It is now generally accepted that within the bounds of distributional assumptions, the Markov approach is the most accurate. If the fluctuating environment is not included in the analysis, the transmission system elements can be considered independent and the probabilities and frequencies calculated directly and simply. In the case of independent components, cut set or tie set methods can also be effectively utilized. When, however, the fluctuating environment, stormy and normal weather, is considered, the statistical behavior of the components cannot be regarded independent and the solution of 2^{n+1} linear algebraic equations is required, where n is the number of components.

When the number of components is large, the number of linear equations becomes unmanageable. Methods like state merging, state space truncation, and sequential truncation have been proposed for alleviating this problem and are described in reference 19. The most efficient method for dealing with transmission and distribution systems, involving dependent modes like the fluctuating environment is the Markov Cut Set method [27]. This method is a combination of the cut set and Markov methods. This composite approach consists in decomposing the system by cut sets and then using Markov processes and frequency balancing concepts for the calculation of the terms in the cut set expansion. The Markov process of only the cut set members is considered and, therefore, a limited number of equations need to be solved at a time. A very useful feature of this approach is that both time-specific and steady-state probabilities and frequencies of system failure can be calculated. It is also possible to control and measure the degree of accuracy of the results. The use of this method is illustrated for transmission systems exposed to a 2-state fluctuating environment. The method can, however, be used to deal with dependence due to maintenance outages and common mode failures.

9.5.1 Minimal Cut Set Method

The equations for the steady-state probability and average frequency of system failure are

$$P_f = \sum_i Pr(C_i) - \sum_{i<j} Pr(C_i \cap C_j) + \sum_{i<j<k} Pr(C_i \cap C_j \cap C_k) \cdots$$

$$(-1)^{m-1} Pr(C_1 \cap C_2 \cap \cdots \cap C_m) \tag{9.70}$$

and

$$f_f = \sum_i Pr(C_i)\bar{\mu}_i - \sum_{i<j} Pr(C_i \cap C_j)\bar{\mu}_{i+j}$$

$$+ \sum_{i<j<k} Pr(C_i \cap C_i \cap C_k)\bar{\mu}_{i+j+k} \cdots (-1)^{m-1} Pr(C_1 \cap C_2 \cap \cdots C_m)$$

$$\bar{\mu}_{1+2+\cdots+m} \tag{9.71}$$

when P_f, f_f = probability and frequency that the system has failed
        C_i = cut set i and also the event: all components of C_i are failed
        m = number of cut sets
$Pr(C_i \cap C_j)$ = probability of the components of both C_i and C_j failed
        μ_j = repair rate of component j
$$\bar{\mu}_{i+k+l} = \Sigma \mu_j$$
$$j \in (C_i \cup C_k \cup C_l)$$

The min cut sets can be calculated using Failure Mode and Effect Analysis and for some well-defined reliability block diagrams, specific algorithms are also available. The min cut sets are defined as sets of minimum number of components whose outage will result in loss of continuity for the system. Once the min cut sets have been determined the probability and frequency of system failure can be determined using (9.70) and (9.71). The mean duration of failure state can be determined using,

$$d_f = \frac{P_f}{f_f} . \tag{9.72}$$

For practical applications, $\lambda_j/\mu_j \ll 1$ and the upper bounds [24] to probability and frequency of failure give results very close to the exact values. These upper bounds are

$$P_{fu} = \sum_i Pr(C_i) \tag{9.73}$$

and

$$f_{fu} = \sum_i Pr(C_i)\bar{\mu}_i \tag{9.74}$$

where P_{fu}, f_{fu} are the first upper bounds to probability and frequency of system failure.

The interval in which P_f and f_f lie can be determined by calculating the lower bounds as well,

$$P_{fl} = \sum_i Pr(C_i) - \sum_{i<j} Pr(C_i \cap C_j) \tag{9.75}$$

and

$$f_{fl} = \sum_i Pr(C_i)\bar{\mu}_i - \sum_{i<j} Pr(C_i \cap C_j)\bar{\mu}_{i+j} \tag{9.76}$$

Increasingly closer upper and lower bounds to P_f and f_f can be obtained by the successive addition of odd and even order terms [24]. It should, however, be reiterated that for almost all the practical systems, the component failure rates are much smaller than the repair rates and therefore (9.73) and (9.74) give results very close to the exact results.

Expression 9.70 for the probability of system failure is true whether or not the components are independent. Equation 9.71 for the frequency of system failure also does not depend on the independence of components so long as there is no restriction on the repair of a failed component [25]. This restrictive assumption can also be eliminated by a more general expression for f_f as shown subsequently. The basic quantities to be calculated in (9.70) and (9.71) are the probabilities of cut sets and their intersections. When the components are assumed independent, these quantities can be calculated by the simple product rule. It is this problem that has restricted the application of cut set methods to the systems where components can be assumed independent. A method is proposed in the next section for calculating these probabilities when the components cannot be assumed independent and this method, therefore, significantly extends the capability of the cut set approach to the transmission and distribution systems.

9.5.2 The Markov Cut Set Method

This section describes a method for calculating the probabilities and frequencies of cut sets and their intersections when the components cannot be regarded statistically independent.

Concept of Equivalent Subset. The proposed method depends upon the relationship between a minimal cut set and the equivalent subset of the system state space. In a cut set C_i, having components l and m as members, if l and m fail, the system will be failed irrespective of the states of the other components of the system. The failure of the members of C_i is equivalent to the system being in subset S_i of the state space S, where

$$S_i \equiv \{s_j: \text{ in the state } s_j, \text{ the components } l \text{ and } m \text{ are failed} \\ \text{ and the other components exist in either state}\}$$

The state s_i^v in which members of C_i are failed and all the other components are good is called the vertex state of subset S_i. The system can transit from the vertex state either upward (in the sense of less components in the failed state) by the repair of the members of C_i or it can transit downward (in the sense of more components failed) by the successive failures of the nonmembers of C_i. The subset S_i consists of states generated by the downward transitions from s_i^v.

It can be seen from the discussion that

$$Pr(C_i) = Pr(S_i) \tag{9.77}$$

where $Pr(S_i) = $ probability of the system being in S_i

$$= \sum_{j:\, s_j \in S_i} Pr(s_j) \qquad\qquad (9.78)$$

where $Pr(s_j) = $ probability of being in system state j.

The frequency formula, not dependent on component independence, can also now be stated in terms of S_i,

$$f_f = \sum_i F(S_i) - \sum_{i<j} F(S_i \cap S_j)$$

$$+ \sum_{i<j<k} F(S_i \cap S_j \cap S_k) - \cdots$$

$$(-1)^{m-1} F(S_1 \cap S_2 \cap \cdots \cap S_m) \qquad\qquad (9.79)$$

where $F(S_i) = $ frequency of encountering subset S_i.

Equation 9.79 is true whether or not the components are statistically independent.

The Method.   The problem of calculating the probabilities and frequencies of cut sets and their intersection can be transformed into that of determining these values for the corresponding equivalent subsets using (9.77)–(9.79). It is now proposed that the Markov and frequency balancing approach be used for the calculation of probabilities and frequencies of equivalent subsets.

Consider, for example, subset S_i equivalent of cut set C_i. If $Pr(S_i)$ and $F(S_i)$ could be calculated from transition rate matrix of only those components that are members of C_i, this would present a big step forward. This is because in a large network, the number of elements in a minimal cut set is generally much smaller than the number of components in the whole network. The number of components to be dealt with at a time can also be kept within reasonable limits by excluding minimal cut sets beyond a specified order. As an example assume that the number of components in a system is 50. The total number of states, if the equations for the entire system are to be solved is 2^{51}, that is, 225×10^{13} when the 50 components are exposed to the same 2-state fluctuating environment. Now if the largest cut set to be considered is of the order 5, then only $2^{5+1} = 64$ states need to be dealt at a time using the purposed method. The solution of 64 linear equations on a modern digital computer is a trivial task.

It can be shown that in a system exposed to a 2-state fluctuating environment $Pr(S_i)$ and $F(S_i)$ can indeed be calculated by developing the transition rate matrix for the elements of C_i only and neglecting other components of the system.

Let n_a be the number of components comprising C_i and $n_b = n - n_a$ be the remaining components, that is, components that are not members of C_i. Here n is the total number of components. The sets of components n_a and n_b will be denoted by X_a and X_b, respectively. Now X_a can exist in 2^{n_a} number of states in each of the environmental states. The component configuration of each state in either of the two environments is the same but the interstate transition rates in the two weather states are different. For the state space corresponding to X_a the transition rate from configuration i to j will be denoted by α_{ij} and β_{ij} in the normal and adverse weather, respectively. The transition rate from a state in the normal weather condition to the corresponding state in the adverse weather condition is assumed as w_n and in the reverse direction as w_m. The states associated with X_a are indicated by $a_1, a_2, \ldots, a_p$ in the normal weather and $a_1^s, a_2^s, \ldots, a_p^s$ in the adverse weather condition p being equal to 2^{n_a}. Similarly the states generated by X_b, that is, components not member of C_i, are indicated by $b_1, b_2, \ldots, b_q$ in the normal weather condition and $b_1^s, b_2^s, \ldots, b_q^s$ in the adverse weather condition, q being equal to 2^{n_b}.

The state space of the entire system can now be generated from the state space of X_a and X_b. When X_a and X_b are combined, there will be $p \times q$ states in each weather condition. This combination of state s for the normal and adverse weather is shown below.

Normal Weather State Space

$$
\begin{array}{cccc}
b_1 a_1 & b_1 a_2 & \cdots & b_1 a_p \\
b_2 a_1 & b_2 a_2 & \cdots & b_2 a_p \\
\vdots & \vdots & \vdots & \\
b_q a_1 & b_q a_2 & \cdots & b_q a_p
\end{array}
$$

Adverse Weather State Space

$$
\begin{array}{cccc}
b_1^s a_1^s, & b_1^s a_2^s, \ldots, & b_1^s a_p^s \\
b_2^s a_1^s, & b_2^s a_2^s, \ldots, & b_2^s a_p^s \\
\vdots & \vdots & \vdots \\
b_q^s a_1^s, & b_q^s a_2^s, \ldots, & b_q^s a_p^s
\end{array}
$$

The transition rate from $b_i a_j$ to $b_i a_k$ is α_{jk}, that is, the transition rate from a_j to a_k. The transition rate from $b_i a_j$ to $b_i^s a_j^s$ is $w_n = 1/T_n$ where T_n is the mean duration of the normal weather. The transition rate from $b_i^s a_j^s$ to $b_i a_j$ is likewise $w_m = 1/T_s$ where T_s is the mean duration of the adverse weather.

Let the states be grouped into subsets D_i and D_i^s in the normal and adverse weather conditions, $i = 1, 2, \ldots, p$. These subsets are such that

$$
D_i = \{b_1 a_i, b_2 a_i, \ldots, b_q a_i\}
$$

and

$$D_i^s = \{ b_1^s a_i^s, b_2^s a_i^s, \ldots, b_q^s a_i^s \}$$

The necessary and sufficient condition [23] for the states to be mergeable into subsets is that for any two subsets D_i, D_j, the transition rate from each state in subset D_i to each of the states in D_j when summed over all states in D_j is the same for each state in D_i and this equivalent transition rate from D_i to D_j is given by

$$\sum_{j \in D_j} \lambda_{ij}$$

When this condition of mergeability is satisfied between all the subsets taken in pairs, the Markov process of the system is said to be mergeable into these disjoint subsets. Let us apply this condition of mergeability to D_i and D_i^s, $i = 1, \ldots, p$.

1. For any two D_i, D_j, $\sum_{j \in D_j} \lambda_{ij}$ is the same for each $i \in D_i$ and is equal to α_{ij} which is the normal weather transition rate from a_i to a_j. Therefore the condition of mergeability is satisfied for any D_i, D_j and the transition rate from subset D_i to D_j is α_{ij}, that is, the transition rate from a_i to a_j.

2. In a similar manner the condition of mergeability is satisfied between any two D_i^s and D_j^s and the transition rate from D_i^s to D_j^s is β_{ij}, that is, the adverse weather transition rate from a_i^s to a_j^s.

3. For any pair D_i and D_i^s, the condition of mergeability is also satisfied and the transition rate from D_i to D_i^s is w_n and from D_i^s to D_i is w_m.

From the preceding discussions, it can be concluded that the Markov process for the system is mergeable into subsets D_i, D_i^s. It can also be recognized that the merged Markov process is identical to the Markov process for components of X_a, that is, components member of cut set C_i. Now if the states a_p and a_p^s represent the failure of all the elements of C_i in the normal and adverse weather respectively, then

$D_p = \{$subset of states in the normal weather having members of C_i failed and other components of the system in either failed or good state$\}$

$D_p^s = \{$subset of states in the adverse weather having members of C_i failed and other components of the system in either of states$\}$

That is,

$$S_i = D_p \cup D_p^s = D_p + D_p^s$$

$$D_p \equiv a_p, \quad D_p^s \equiv a_p^s$$

The merged process is identical to the process corresponding to the members of cut set C_i. Therefore

$$Pr(C_i) = Pr(S_i)$$

$$= Pr(D_p) + Pr(D_p^s)$$

$$= Pr(a_p) + Pr(a_p^s) \tag{9.80}$$

$$F(S_i) = F(D_p \cup D_p^s)$$

$$= F(a_p \cup a_p^s) \tag{9.81}$$

Here $F(a_p \cup a_p^s)$ is the frequency of encountering the state where all the elements of C_i are failed and can be readily calculated using frequency balancing concept. For steady-state

$$F(a_p \cup a_p^s) = \sum_{i \neq p} \left[Pr(a_p)\alpha_{pi} + Pr(a_p^s)\beta_{pi} \right] \tag{9.82}$$

From (9.80) and (9.81) it can be seen that the probability and frequency of a cut set C_i, for the system exposed to fluctuating environment, can be calculated by considering the Markov process associated with the members of cut set C_i only and that the transition rate matrix of the entire system need not be generated. Therefore, the terms in (9.70) and (9.71) can be computed by generating the transition rate matrix of the elements of each cut set or intersection at a time and as noted previously these matrices are much smaller in size than the matrix for the entire system. It is to be noted that since the necessary and sufficient condition of mergeability is satisfied, (9.80) and (9.81) can be used for both time specific and steady state.

It becomes, however, obvious that the higher the order of intersections considered, the less advantageous the procedure becomes since the number of components to be considered at a time increases. Therefore, for the successful implementation, the following procedure is suggested.

PROCEDURE

1. Identify the cut sets to be considered. The cut sets having more than x components may be ignored. It will be reasonable to have $x=5$, since the probability of more than 5 overlapping outages can safely be regarded negligible.
2. Since in all practical systems the component failure rate is much smaller than the repair rate, the upper bound will give an almost exact result. Therefore P_f and f_f can be approximated as

$$P_f \simeq P_{fu} = \sum_i Pr(C_i) \tag{9.83}$$

and

$$f_f \simeq f_{fu} = \sum_i F(S_i) \qquad (9.84)$$

3. The terms of (9.83) and (9.84) can be calculated by generating the transition rate matrix of the components of each cut set at a time and computing $Pr(C_i)$ and $F(S_i)$ using (9.80) and (9.81).

It can be seen that if the above procedure is followed only 2^{x+1} equations need be solved at a time, x being the number of elements in cut set. If $x=5$, it means 64 equations, which is a trivial task when digital computers are employed. The calculation of the first lower bound will not involve much additional difficulty and can provide insight on the margin of error.

It should be carefully noted that (9.80) and (9.81) for the calculation of the terms of (9.70) and (9.71) or (9.83) and (9.84) are exact. The approximation involved is either in the ignoring of higher order cut sets or using upper bound approximations by (9.83) and (9.84) instead of complete (9.70) and (9.71).

Comparison with State Space Truncation.   If the cut sets of say order 6 or higher were to be ignored, one might ask, "How is the Markov cut set approach superior to state space truncation when contingencies of order higher than 5 are ignored?" Consider a system of say n components exposed to normal and adverse weather. If contingencies of the order higher than x are ignored, the number of linear equations involved is,

$$= 2 \sum_{i=0}^{x} \binom{n}{i}$$

If, for example, $n=50$ and $x=5$, the total number of states or corresponding equations is 2369936, which is beyond the capability of the today computers. On the other hand using Markov cut set approach only 64 equations need be considered simultaneously, which for state space truncation, corresponds to considering only single-order contingencies.

APPROXIMATIONS AND EXTENSIONS.   The Markov cut set method has been shown to deal in an exact manner with the form of dependency induced by the fluctuating environment. Even though this method may not exactly apply to all forms of dependency, it could provide good approximations for certain limited forms of component dependence due to maintenance outages and common-mode failures.

Example.   The Markov Cut-set method is illustrated by application to a complex configuration shown in Figure 9.16. The cut sets of this system are identified in Table 9.3. The relevant data is shown in Table 9.4, which

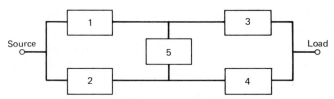

Figure 9.16 An example of a complex configuration.

also gives the comparison of the results by exact Markov and Markov cut set methods. The results obtained by the Markov cut set method are only negligibly higher than the Markov method. The mean duration of the failure state can be obtained by the (9.72). The error is somewhat higher for higher percentage of failures during adverse weather. This is to be expected since this results in effectively increasing the ratio of failure rate to repair rate. It should be reiterated here that any error introduced by the Markov cut set method is because of the use of upper bound approximation and not in the calculation of terms of the equations for the probability and frequency of failure.

Table 9.3 *Minimal Cut Sets of Figure 9.16*

Cut Set	Component Members
C_1	1,2
C_2	3,4
C_3	1,4,5
C_4	2,3,5

Table 9.4 *Comparison of Markov and Markov Cut Set Methods.*
Components are assumed identical. Average failure rate = 0.5 failure/year. Normal weather mean duration = 200 hours. Adverse weather mean duration = 1.5 hours.

Mean Down Time of Each Component (hours)	Failures During Adverse Weather, %	Failure Probability		Failure Frequency (per year)	
		Markov	Markov Cut Set	Markov	Markov Cut Set
5	20	3.4996×10^{-7}	3.4996×10^{-7}	1.2275×10^{-3}	1.2275×10^{-3}
5	80	3.3144×10^{-6}	3.3157×10^{-6}	1.1674×10^{-2}	1.1684×10^{-2}
10	20	1.0736×10^{-6}	1.0736×10^{-6}	1.8830×10^{-3}	1.8831×10^{-3}
10	80	7.7573×10^{-6}	7.7633×10^{-6}	1.3678×10^{-2}	1.3695×10^{-2}

ON QUALITY OF SERVICE. The method discussed can be used for both continuity of service criteria as well as indices considering the voltage levels. The difference lies in the calculation of minimal cut sets. The process proceeds in essentially two steps. First a set of components is assumed out of service due to forced outage or maintenance outage. Given this event, the level that will cause unacceptable voltage level is then determined using load flow. Repetition of this procedure for different sets of component outages, then identifies the minimal cut sets in terms of component outages and load levels. Once the minimal cut sets have been identified, the Markov cut set method can be used.

REFERENCES

1. Allen, R. N., and R. A. F. Nunes, "Modelling of Stand-By Generating Units in Short-Term Reliability Evaluation," *IEEE Power Engineering Society Winter Meeting*, Paper No. A 79 006-8, New York, Feb. 1979.

2. Anstine, L. T., R. E. Burke, J. E. Casey, R. Holgate, R. S. John, and H. G. Stewart, "Application of Probability Methods to the Determination of Spinning Reserve Requirements for the Pennsylvania-New Jersey-Maryland Interconnection," *IEEE Trans. Power Appar. Syst.*, **PAS-82**, 720–735 (1963).

3. Billinton, R. and C. Singh, "Generating Capacity Reliability Evaluation in Interconnected Systems Using A Frequency and Duration Approach—Part I—Mathematical Analysis," *IEEE Trans. Power Appar. Syst.*, **PAS-90**, 1646–1654 (1971).

4 Billinton, R. and C. Singh, "Generating Capacity Reliability Evaluation in Interconnected Systems Using a Frequency and Duration Approach—Part II—System Applications," *IEEE Trans. Power Appar. Syst.*, **PAS-90**, 1654–1669 (1971).

5. Billinton, R. and C. Singh, "Generating Capacity Reliability Evaluation in Interconnected Systems Using a Frequency and Duration Approach—Part III—Correlated Load Models," *IEEE Trans. Power Appar. Syst.*, **PAS-91**, 2143–2153 (1972).

6. Billinton, R. and C. Singh, "Generating Capacity Reliability Evaluation," *Canadian Electrical Association*, Spring Meeting, March 1971, Vancouver.

7. Billinton, R. and C. Singh, "System Load Representation in Generating Capacity Reliability Studies—Part I—Model Formulation and Analysis," *IEEE Trans. Power Appar. Syst.*, **PAS-91**, 2125–2132 (1972).

8. Billinton, R. and C. Singh, "System Load Representation in Generating Capacity Reliability Studies—Part II—Applications and Extensions," *IEEE Trans. Power Appar. Syst.*, **PAS-91**, 2133–2143 (1972).

9. Billinton, R., *Power System Reliability Evaluation*, Gordon and Breach, New York, 1970.

10. Billinton, R. and A. V. Jain, "The Effect of Rapid Start and Hot Reserve Units in Spinning Reserve Studies," *IEEE Trans. Power Appar. Syst.*, **PAS-91**, 511–516 (1972).

11. Endrenyi, J., *Reliability Modeling in Electric Power Systems*, Wiley, New York, 1978.

12. IEEE Report, "Bibliography on the Application of Probability Methods in Power System Reliability Evaluation," *IEEE Trans. Power Appar. Syst.*, **PAS-97**, 2235–2242 (1978).

13. IEEE Committee Report, "Definitions of Customer and Load Reliability Indices for Evaluating Electric Power System Performance," IEEE Paper A 75 588-4, PES Summer Meeting, San Francisco, CA, July 1975.

14. Patton, A. D., "Short-Term Reliability Calculation," *IEEE Trans. Power Appar. Syst.*, **PAS-89**, 509–514 (1970).

15. Patton, A. D., "Assessment of the Security of Operating Electric Power Systems Using Probability Methods," *Proc. IEEE*, **62**, 892–901 (1974).

16. Ringlee, R. J. and A. J. Wood, "Frequency and Duration Methods for Power System Reliability Calculations—Part II—Demand Model and Capacity Reserve Model," *IEEE Trans. Power Appar. Syst.*, **PAS-88**, 375–388 (1969).

17. Singh, C., "Generating Capacity Reliability Evaluation Using Frequency and Duration Methods," MSc. Thesis, University of Saskatchewan, 1970.

18. Singh, C., "Reliability Modelling and Evaluation in Electric Power Systems," Ph.D. Thesis, University of Saskatchewan, 1972.

19. Singh, C., "Reliability Calculations of Large Systems," *Proceedings of the Annual Reliability and Maintainability Symposium*, Washington, D.C., IEEE, New York, 1975.

20. Singh, C., "Reliability Modeling Algorithms For a Class of Large Repairable Systems," *Microelectron. Reliability*, **15**, 159–162, 1976.

21. Singh, C., "Computer Program For Generating Capacity Reliability Evaluation Using Frequency and Duration Methods," information available from author.

22. Singh, C. and R. Billinton, "A Frequency and Duration Approach to Short-Term Reliability Evaluation," *IEEE Trans. Power Appar. Syst.*, **PAS-92**, 2073–2082 (1973).

23. Singh, C. and R. Billinton, *System Reliability Modelling and Evaluation*, Hutchinson, London, 1977.

24. Singh, C., "On the Behaviour of Failure Frequency Bounds," *IEEE Trans. Reliab.*, **R-26**, 63–66 (1977).

25. Singh, C., "On s-Independence in a New Method to Determine the Failure Frequency of a Complex System," *IEEE Trans. Reliab.*, **R-27**, 147–148 (1978).

26. Singh, C., "Effect of Probability Distributions on Steady State Frequency," *IEEE Trans. Reliab.*, **R-29**, (1980).

27. Singh, C., "Markov Cut-Set Approach for the Reliability Evaluation of Transmission and Distribution Systems," IEEE Paper A 80 069-5, IEEE PES Winter Meeting, New York, Feb. 1980.

10

Transit System Reliability

10.1 INTRODUCTION

Reliability is an important consideration in the planning, design, and operation of transit systems. The discussion in this chapter is focused on track bound transit systems; the principles can, however, be applied to other types of transit systems as well. The term track bound is used here to describe systems whose vehicles are captive on a common track. This includes steel wheel on steel rail, rubber wheel on concrete guideway, and magnetically levitated vehicles. In the case of a road system, the failure of a vehicle affects the concerned vehicle and some delay may be caused to the other vehicles. The effect on the system is, however, more or less localized since the failed vehicle can be pulled to the side or bypassed by the other vehicles. The bypass capability of the track bound systems, on the other hand is extremely limited. The failure of a single vehicle in such systems could affect or immobilize the upstream vehicles and depending upon network configuration, the degrading or immobilizing effect could spread over the entire or a major part of the system. This serial effect makes the reliability an all the more important consideration in track bound transit systems.

Reliability is important for both the transit operator and the passengers. Lower reliability means increased unscheduled maintenance and decreased equipment availability. If availability is low, more vehicles are needed to meet the passenger demand but even with more vehicles, system performance may not be satisfactory. More vehicles can increase system availability but do not decrease the incidence of system failures. Reliability is important to passengers as it reflects the ability of a transit system to keep operating schedules.

Traditionally, the transit operators have been relying on warranties to assure the procurement of reliable equipment. The warranties are, however, more like maintenance or service contracts and do not necessarily serve as deterrents to system unreliability. Warranty makes the manufacturer pay for repairs during a limited period of time but once the warranty period is over, unreliability becomes the headache of the transit operator. The operators are now realizing that economical reliability can be built into the systems only during the design, development, and manufacturing stage of the equipment. The emphasis now appears to be shifting towards specifying reliability targets and implementing reliability assurance programs during design and development. Reliability models provide a useful tool for the analysis of design configurations. These models for the components of a transit system and the techniques of combining these models are described in this chapter.

The models are discussed with regard to a specific system configuration in which the track is a single loop (Figure 10.1) with off-line stations. From the view point of reliability modeling, this configuration can be generally regarded equivalent to a single two way route in which the vehicles move on one track in one direction and at the terminal station they are switched on to the other track for the opposite direction. The methodology illustrated by these models can be employed to develop models for more individual applications.

A transit system consists of the following subsystems:

1. Vehicle fleet.
2. Passenger stations.
3. Substations for power supply.
4. Command and control.
5. Guideway.

These subsystems, in turn, may be divided into what is termed components in this chapter. Models for each subsystem are described.

10.2 BASIC THEORY

The models consider the total and partial failure modes of the vehicle, the spare vehicle, and vehicles on maintenance. These models are then combined with the models for the other subsystems. There is a considerable statistical dependence between the various modes of failure, repair, retrieval, and maintenance. The state space approach [9, 10] is, therefore, used in developing these models.

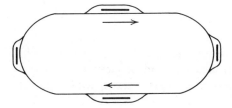

Figure 10.1 A single loop.

10.2.1 State Space Approach

In this approach, the possible states of the system (state space) and the modes of transiting from one state to another are identified. Each mode of interstate transition is assigned a specific value called its interstate transition rate. The state equations can be written using the frequency balancing approach [10]. The frequency of transiting from state i to state j is defined as the expected transition rate from state i to state j and is given by $P_i \lambda_{ij}$. The frequency balancing approach states that the rate of change of the probability of being in state i equals the frequency of transiting into state i from all the remaining states minus the frequency of transiting out of state i, that is,

$$\sum_j P_j(t)\lambda_{ji} - P_i(t)\sum_j \lambda_{ij} = \dot{P}_i(t) \tag{10.1}$$

In the equilibrium condition, $\dot{P}_i(t)=0$, and therefore (10.1) reduces to

$$\sum_j P_j \lambda_{ji} - P_i \sum_j \lambda_{ij} = 0 \tag{10.2}$$

that is, the frequency of encountering state i equals the frequency of encountering the rest of state space from state i. For n states there are n equations and they are linearly dependent; any equation can be obtained from the remaining $(n-1)$ equations. Any $(n-1)$ equations together with the total probability equation,

$$\sum P_i = 1$$

can be solved to obtain the state probabilities. These state probabilities can be used to obtain the reliability measures.

The number of states tend to be large due to the size and complexity of the transit system. The models can be reduced using the concept of equivalent transition rate [10], which under the equilibrium condition is given by

$\lambda^e_{X^- - X^+}$ = the equivalent transition rate from subset X^- to subset X^+

$$= \sum_{i \in X^-} \sum_{j \in X^+} \frac{P_i \lambda_{ij}}{\sum_{i \in X^-} P_i} \tag{10.3}$$

The other techniques used for keeping the number of states within manageable limits are state space truncation and sequential truncation [10]. These techniques systematically omit or delete states with relatively low probabilities.

10.2.2 Measures of Reliability

Transit systems, like other commercial or public systems, are designed to meet a certain demand. Therefore, the reliability of transit systems can be viewed in two ways. The system is comprised of hardware, software, and the human interface, although in most of the reliability studies only hardware and computer-based software (if any) are considered. One way of looking at the reliability is in terms of the system deficiencies. The measures relating to this approach are termed system-based reliability measures. It is also desirable to know how these system deficiencies relate to the inability of the system to satisfy the demand and the corresponding measures are called the demand-based reliability measures. Obviously both of these types of measures are interrelated.

System-Based Measures. Reliability indices are usually defined in terms of success or failure. Many complex systems like transit systems or electric power systems have, however, several levels of failure and it is, therefore, appropriate to define the calculated reliability measures in terms of subset X^+, which may contain a specific number of system states. This subset defines an event or a particular mode of degradation of the system. The various modes or levels of system degradation can, therefore, be represented by suitably defining the elements of X^+. As an example X^+ may be used to represent the system states having the number of failed passenger stations greater than a particular number. The following measures defined on X^+ have been used in this chapter.

1. *Probability of X^+.* This can be defined as the limiting value of the time spent in X^+ as a fraction of the total operating time and is given by

$$P_+ = \sum_{i \in X^+} P_i \tag{10.4}$$

2. *Frequency of encountering X^+.* This is the mean number of occurrences of X^+ per unit of the operating time and can be calculated by

$$f_+ = \sum_{j \in X^-} P_j \sum_{i \in X^+} \lambda_{ji} \tag{10.5}$$

3. *Mean cycle time of X^+.* This is the mean time between successive encounters of X^+ and equals the reciprocal of f_+, that is,

$$T_+ = \frac{1}{f_+} \tag{10.6}$$

4. *Mean duration of* X^+. This is the expected time of stay in X^+ in one cycle (one cycle constitutes X^+ and X^-) and is given by

$$d_+ = \frac{P_+}{f_+}$$

$$= P_+ T_+ \tag{10.7}$$

Demand Based Measures [*13*]. The primary purpose of a transit system is to move passengers between the various points of a network. It is, therefore, important to have a measure of reliability as perceived by the passengers. As an example [7] consider a jeep having an MTBF of approximately 260 hours [5]. If this vehicle were driven on the average for 30 miles/day at an average speed of 15 miles/hour the average interval between two failures would be 130 calendar days, whereas this interval in terms of operating time would be only about 11 days. Coming back to transit systems, consider a system that breaks down on an average of every 15 calendar days with the average down time duration of half an hour. A passenger who travels for only, say, 15 minutes twice a day will not be affected by every system failure. Assuming uniform service level of 16 hours/day operation, the passenger may be affected on the average approximately by every seventh failure and therefore will tend to see the system failing on the average approximately three times a year. The perception of the failure is further affected by several factors such as whether the delay has to be tolerated in a comfortable, airconditioned environment or in a hot stuffy vehicle and the personal temperament of the passenger. Media reports and the stories of system failures told by other passengers add something to the direct exposure to failures. It can be appreciated that it is extremely difficult to measure or predict the passengers perception of the failures. Nevertheless suitable measures related to the impact of system failures on the passengers can be devised.

The ability of a transit system to continue providing transportation services as scheduled or advertised may be termed as the operational reliability. In this definition, it is assumed that schedules are set within normal capabilities of the system. If the transit system cannot provide scheduled service when every subsystem is working normally, it is a problem of planning, scheduling, or operations management. A failure in a subsystem, however, causes a perturbation that may affect the system's ability to provide adequate service. The impact of failures on the schedule-keeping ability of a transit system is the concern of operational reliability. The demand based measures of reliability are, therefore, also the measures of operational reliability. Some measures are described here.

DELAY, THE BASIC MEASURE OF OPERATIONAL RELIABILITY [11]. So long as a passenger can get from one station to another in a comfortable, safe, and timely manner, a failure occurring in a system does not bother him. A minor delay may be hardly noticed, but if the passengers suffer long delays or if the delay is too frequent, the transit system would appear unreliable to the passengers. The failure-induced delay is, therefore, a meaningful measure of operational reliability. The delay may be incurred at the following points.

Departure points. When a passenger arrives at a departure point, the system may be inoperative or operative in a degraded mode. This adds to the waiting time, making the total travel time longer.

Delay during travel. The passengers on board the vehicles may suffer delay due to the breakdown of a subsystem. The delay can be broadly classified into three categories:

1. Minor delay $<x$ units of time.
2. Major delay $>x$ units of time.
3. Entrapment, a major delay requiring passenger evacuation.

There does not appear to be enough relevant data on the passenger intolerance of delays and this appears to be a useful although difficult area for investigation. Despite extensive investigations, some judgment will always be involved in fixing the value of x. Some basic quantitative measures of operational reliability in terms of delay are discussed below [11].

1. $P_D(x_1, x_2)$, that is, the probability that a passenger will encounter a delay, $>x_1$ and $<x_2$, on a trip. The length of the delay is defined by the interval (x_1, x_2). The interval $(0, x)$ means a delay less than x, whereas (x, ∞) means delay longer than x.
2. MTBD, the mean time between two successive delays suffered by a passenger.
3. Expected value of delay.

Using the frequency concept of probability, $P_D(x_1, x_2)$ can also be interpreted as the limiting value of the proportion of the trips having a delay (x_1, x_2) to the total number of trips. Suppose that a person makes a large number of trips of varying lengths. If the trips on which he incurred a delay of say greater than 5 minutes were counted and then divided by the total number of trips, it would approximate $P_D(5, \infty)$. The probability can be further converted into MTBD by knowing the number of trips in a year. It will also be desirable to compute the expected value of delay. Data, however, may not be available to compute these measures. The calculation of these measures could be simplified by relating the delay to the vehicles rather than the passengers. In such a case the ratio of vehicle delayed trips to the total trips would be calculated.

Many measures of operational reliability can be defined [2–4]. These measures should reflect the delay incurred or travel time lost by the passengers or vehicles. The real difficulties lie, however, not in defining measures of operational reliability but in developing suitable analysis techniques and obtaining valid data for calculating these measures.

LOCE OR LOCP.   As noted earlier it is not difficult to define more sophisticated measures of operational reliability. The harder part is the data and subsequent synthesis of this data to calculate the measure. Any measure may prove to be satisfactory so long as it reasonably reflects the ability of the system to provide adequate transportation service. One simpler measure may be called "Loss of Capacity Expectation" (LOCE) or "Loss of Capacity Probability" (LOCP) and can be defined as the probability that the system will not have enough capacity to meet the demand adequately. This can also be interpreted as the expected value of time during which the system cannot meet the demand. This will include the periods of degraded operation, for example, not enough vehicles being available or some other subsystem failure causing deficient operation. The LOCE is computed as

$$\text{LOCE} = \sum_i P_i Q_i \tag{10.8}$$

where P_i = probability of the system being in state i
        Q_i = probability that the system will not be able to meet demand in state i

The LOCE gives the same weight to all system deficiencies irrespective of the magnitude of the impact and it is therefore likely to be a conservative measure. This approach is identical to the use of LOLE (loss of load expectation) in power generation planning studies [1] by the electric power utilities.

10.3 VEHICLE SYSTEM MODEL

This section first describes the model for a single vehicle and then for the system of vehicles. The vehicle includes the body structure and all onboard equipment carried by the vehicle. A vehicle can have several modes of failure. For the sake of simplicity, however, each component of the vehicle is assumed to have two modes of failure.

Retrieval or Total Failure Mode.   With this type of failure, the vehicle is immobilized on the track and it cannot move on its own. External assistance is normally required for clearing it from the track. This kind of failure is severe and causes serious delay to the passengers as not only the affected vehicle is stuck but the upstream vehicles also come to a halt.

Partial Failure Mode. This type of failure causes degradation of vehicle performance but the vehicle is not immobilized and can clear the track on its own. Such failures cause system slow down but do not interrupt the service. When the defective vehicle clears the track by getting into a siding or into the maintenance yard, the normal flow of traffic is resumed.

10.3.1 Vehicle Model

The reliability model for a single track bound transit vehicle is shown in Figure 10.2, where the following notation is used:

$\lambda_{i1}, \lambda_{i2}$ = the retrieval and partial mode failure rates of the ith component, that is,

λ_{i1} = 1/(mean time between retrieval mode failures of the ith component)

λ_{i2} = 1/(mean time between partial mode failures of the ith component)

μ_r = retrieval rate = 1/mean time to retrieve a vehicle

μ_p = partial failure recovery rate = 1/mean time to clear a partially failed vehicle

μ_i = repair rate of the ith component = 1/mean time to repair the ith component

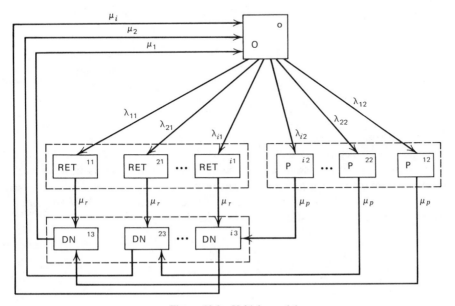

Figure 10.2 Vehicle model.

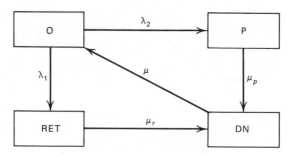

Figure 10.3   Equivalent vehicle model.

The normally operating state of the vehicle is denoted by "O." From this state, the vehicle can fail into a retrieval mode failure (denoted by RET) or a partial mode failure, denoted by P. The vehicle is either retrieved from the track or in the case of partial failure it clears the track on its own power and passes into the down state (denoted DN) in the maintenance area. The vehicle is then repaired back into its normally operating state O. It should be noted that the mean retrieval time is assumed to be the same for failures originating from different components. The same is true for the mean time to clear a partial mode failure.

The vehicle states can be grouped using the concept of equivalent transition rate defined in (10.3). The equivalent model is shown in Figure 10.3 where

Equivalent state (Figure 10.3)	Original states (Figure 10.2)	Description
O	O	operating state
RET	$(11, 21, \ldots, i1, \ldots)$	retrieval mode failure
P	$(12, 22, \ldots, i2, \ldots)$	partial mode failure
DN	$(13, 23, \ldots, i3, \ldots)$	being repaired off track

Strictly speaking, the merging of states $(13, 23, \ldots, i3, \ldots)$ is correct only if the component repair rates are equal (see conditions of mergeability [10]). The error introduced because of nonequality of repair rates is, however, relatively small. The equivalent transition rates of Figure 10.3 can be calculated using (10.3). As an example, for calculating λ_1, the equivalent transition rate from the state O to state RET,

$$X^- = (O)$$

and

$$X^+ = (11, 21, \ldots, i1, \ldots)$$

and therefore

$$\lambda_1 = \sum_i \frac{P_0 \lambda_{i1}}{P_0}$$

$$= \sum_i \lambda_{i1} \tag{10.9}$$

Similarly for μ, the equivalent transition rate from the state DN to the state O,

$$X^- = (13, 23, \ldots, i3, \ldots)$$

and

$$X^+ = (O)$$

Therefore

$$\mu = \frac{\sum_i P_{i3} \mu_i}{\sum_i P_{i3}} \tag{10.10}$$

Now

$$P_{i3} \cdot \mu_i = P_{i1} \mu_r + P_{i2} \mu_p$$

$$= P_0 \lambda_{i1} + P_0 \lambda_{i2}$$

that is,

$$P_{i3} = \frac{(\lambda_{i1} + \lambda_{i2}) P_0}{\mu_i} \tag{10.11}$$

Substituting P_{i3} from (10.11) into (10.10)

$$\mu = \frac{\sum_i (\lambda_{i1} + \lambda_{i2})}{\sum_i (\lambda_{i1} + \lambda_{i2}) T_i} \tag{10.12}$$

where

$$T_i = \frac{1}{\mu_i}.$$

In a similar fashion it can be proved that the equivalent transition rates μ_r and μ_p in Figure 10.3 have the same values as μ_r and μ_p in Figure 10.2.

10.3.2 Vehicle System Model

The state space model of a single vehicle is described in section 10.3.1 and this section now describes the model for the system of vehicles, that is, all the passenger carrying vehicles in the system. The model is based on the following assumptions:

1. Only the operating vehicles are liable to fail and the vehicles on standby or maintenance are not subject to failure. This assumption results in assigning zero failure rates to the vehicles being maintained or in standby mode. The failure rate of a cold standby can be generally assumed zero. If the vehicles are in a warm standby mode, that is, partially powered, this assumption is valid only so long as the failure rate in warm standby mode is small as compared with that of the operating vehicle.

2. The failure of a single vehicle in the retrieval mode causes the whole system of vehicles to be down. The duration of the down time of a vehicle is considered from the time it comes to a halt to time of resuming normal operation. The down time of all the vehicles, the directly affected vehicle and the vehicles coming to a halt as a result of blocking, is assumed to be the same. This assumption was made because of short headways and a relatively small loop length. In a larger loop and longer headways, all the vehicles may not be equally affected and a correction to models may be needed.

3. A vehicle in a partial failure mode is assumed to be removed from the system as soon as possible after the occurrence of the failure and therefore the probability of another unit failing during this period or the partial mode passing into full mode is assumed zero.

4. So long as there is even one standby, a unit on which maintenance is completed will be interchanged with an operating or standby unit. When, however, no spares are left, the unit passes directly from maintenance into the operating mode, without going on stand-by.

A section of the state transition diagram of the model for the system of vehicles is shown in Figure 10.4, where n, s, and m denote the number of operating, spare, and on-maintenance (preventive) units, respectively. The bigger squares represent the operating states and the small squares denote the corresponding partial and retrieval mode states. The index in the top left corner of the operating state is the state number and the number of failed units is indicated in the lower left corner. The first column of states has m units on maintenance and is called group m in Figure 10.4, the second column is called group $(m-1)$ and has $(m-1)$ units on maintenance and so on. Only two groups are shown in Figure 10.4. In the state

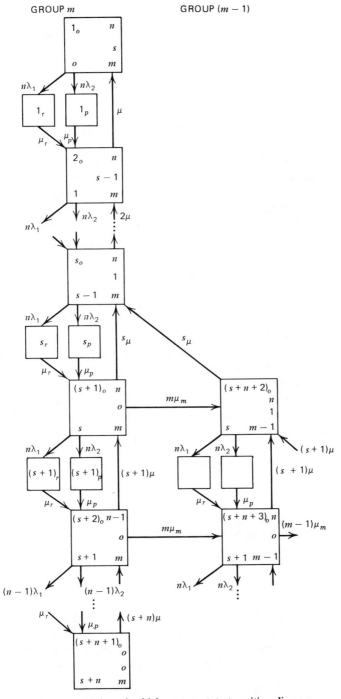

Figure 10.4 A section of vehicle system state transition diagram.

index i_x, i denotes the state number and x denotes the affiliation

$o =$ operating state
$r =$ retrieval state
$p =$ for partially operating state

For example, 2_o indicates state 2 is in the operating mode and 2_r and 2_p indicate the associated retrieval and partial modes. The relationships between the various states can be better understood by tracing through a few of the states. Starting at the top, 1_o represents the state when every unit is as it should be. In this state there are n operating units, s spare units to replace the failed units and preventive maintenance is being carried on m units and there is no failed unit. From state 1_o, the system could transit into state 1_r or 1_p by the failure of a unit in the retrieval and partial mode, respectively. In state 1_r, the failed vehicle sits on the track and brings the system to a halt. After the failed vehicle is removed, the system enters state 2_o, there is one failed unit, and this failed unit has been replaced by a spare unit, reducing the number of spares by one. Similarly from state 1_p, the system will transit to state 2_o by the partial failure recovery. From state 2_o, the system could transit into state 1_o by the repair of the failed unit or it could transit into 2_r or 2_p by the failure of another unit. This pattern of transitions continues until state $(s+1)_o$ with s failed units and o units on stand-by is reached. In state $(s+1)_o$, in addition to the pattern of transitions discussed earlier, another mode of transition is introduced, that is, when maintenance on a unit is now completed, it is put in the standby mode (group $m-1$). Now consider state $(s+2)_o$ in which $(n-1)$ units are operating, that is, one less than the required number. If maintenance is now completed on a unit, it is put into the operating mode and the system transits into $(s+n+3)_o$. The rest of the states can be traced in a similar manner.

10.3.3 Reduced Vehicle System Model

The number of system states can be derived from Figure 10.4 as,

$$NVS = 3(s-1) + (m+1)(3n+4) \qquad (10.13)$$

For example for $n=50$, $s=2$, and $m=4$, $NVS=773$ states. The number of states can be considerably reduced by merging i_r and i_p with i_o. The reduced model is shown in Figure 10.5 and its state i is equivalent to (i_o, i_r, i_p) of Figure 10.4. The number of states is given by

$$RNVS = (s-1) + (m+1)(n+2) \qquad (10.14)$$

Now for $n=50$, $s=2$, and $m=4$, $RNVS=261$ as compared with $NVS=773$. For m units on maintenance, there are $(m+1)$ groups of states of Figure

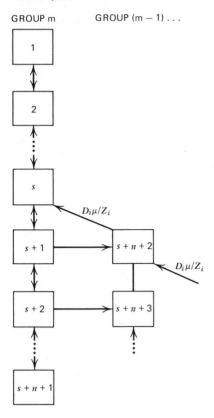

GROUP m   GROUP (m − 1) . . .

Figure 10.5 A section of reduced state transition diagram of vehicle system.

10.5. The equivalent transition rates between various states of the Figure 10.5 can be calculated using (10.3).

Equivalent Transition Rates within a Particular Group. The transition rate from state i to state $i+1$ is

$$\lambda_{i(i+1)} = \frac{P_{ip} \cdot \mu_p + P_{ir} \cdot \mu_r}{P_{io} + P_{ip} + P_{ir}} \tag{10.15}$$

Now

$$P_{ir} = \frac{O_i \lambda_1}{\mu_r} \tag{10.16}$$

and

$$P_{ip} = \frac{O_i \lambda_2}{\mu_p} \tag{10.17}$$

where O_i is the number of operating units in state i. Substituting for P_{ir} and P_{ip} from (10.16) and (10.17) into (10.15),

$$\lambda_{i(i+1)} = \frac{O_i(\lambda_1 + \lambda_2)}{Z_i} \tag{10.18}$$

where

$$Z_i = 1 + O_i\left(\frac{\lambda_1}{\mu_r} + \frac{\lambda_2}{\mu_p}\right)$$

Similarly

$$\lambda_{i(i-1)} = \frac{P_{io}\mu D_i}{P_{io} + P_{ir} + P_{ip}}$$

$$= \frac{\mu D_i}{Z_i} \tag{10.19}$$

where D_i is the number of failed units in the state i.

Equivalent Transition Rates from a Particular Group to the Next Group (say j to $j+1$). The equivalent transition rate from state i to $i+n+1$ is given by

$$\mu_{i(i+n+1)} = \frac{m_i\mu_m I_i P_{io}}{P_{io} + P_{ir} + P_{ip}}$$

$$= \frac{m_i\mu_m I_i}{Z_i} \tag{10.20}$$

where m_i = number of units on maintenance in state i
$\mu_m = 1/T_m$
T_m = mean maintenance time
$I_i = 0$ for number of spares in state $i > 0$
$\quad\ = 1$ otherwise

The Equivalent Transition Rates From a Particular Group to the Previous Group. Shown in Figure 10.5.

10.3.4 Solution for State Probabilities and Reliability Measures

State Equations. The steady-state equations for the equivalent model (Figure 10.5) can be written using (10.2). In the matrix notation this can be written in the form,

$$AP = O \tag{10.21}$$

where $A =$ an $N \times N$ matrix such that its ijth term represents the transition rate from state j to state i, N being the total number of states

$P =$ a column vector whose ith term is P_i, that is, the probability of being in state i

$O =$ is a column vector having all elements as zero

The N equations of (10.21) are linearly dependent, that is, any equation can be obtained from the remaining $(N-1)$ equations. Therefore any $(N-1)$ equations of (10.21) together with the total probability equation (10.22) can be solved to obtain P

$$\sum_i P_i = 1.0 \qquad (10.22)$$

In the matrix form any row of A and the corresponding element in O are changed to 1.0 before solution. The linear equations can be solved using numerical methods like Gauss elimination. Once the probabilities of the equivalent states have been determined, the probabilities of the original states can be calculated using the following equations:

$$P_{io} = \frac{P_i}{Z_i} \qquad (10.23)$$

$$P_{ir} = \frac{O_i \lambda_1 P_{io}}{\mu_r} \qquad (10.24)$$

and

$$P_{ip} = \frac{O_i \lambda_2 P_{io}}{\mu_p} \qquad (10.25)$$

Vehicle Exposure Factor. The vehicle system model described in Section 10.3.2 is good for the on-line stations. When, however, the stations are off-line, the retrieval mode failure of a vehicle within the station limits does not immobilize the rest of the vehicles since they can bypass on the express lane. The retrieval state i_r can, therefore, be decomposed into i_{r1} and i_{r2}, representing failure within the station limit and on-line, respectively. The probabilities for these states can be computed as follows:

$$P_{i_{r1}} = P_{ir} \cdot E \qquad (10.26)$$

and

$$P_{i_{r2}} = P_{ir} \cdot (1 - E) \qquad (10.27)$$

where E is the exposure factor, defined as the limiting ratio of the unit failures within the station limits to the total failures and is assumed approximately equal to the ratio of the length of the guideway within station limits to the total length of the guideway.

The vehicle failures within the station limits will, however, shut down the station lane and are, therefore, considered a part of the station lane failures. The equivalent failure rate component to be added to the station lane failure rate is calculated by,

$$\lambda_{VS} = \frac{\sum\limits_{i} P_{io} O_i \lambda_1 E}{\sum\limits_{i} P_{io} \cdot N_{ST}} \tag{10.28}$$

where N_{ST} is the number of passenger stations.

Vehicle System Reliability Measures. Once the state probabilities have been calculated, the event (subset of states) probabilities, frequencies, and other measures can be calculated using (10.4)–(10.7). There are many ways of defining the events and two of them are described below.

EXACT MEASURES. These measures calculate the probabilities and frequencies of encountering states in which the operating vehicles are equal to a particular number and can be designated as $P(N_o = n)$ and $f(N_o = n)$ where N_o denotes the operating vehicles. The probability of $N_o = n$ can be simply calculated by adding the probabilities of all states having n operating vehicles. The frequency and other measures can be calculated using (10.5)–(10.7).

CUMULATIVE STATE MEASURES. Another way to represent these measures is to calculate the probabilities and frequencies of encountering states in which there are fewer than a particular number of operating vehicles. The equations for these measures are given below.

$$P(N_o \leqslant n) = \sum\limits_{i:\, O_i \leqslant n} (P_{io} + P_{ir1} + P_{ip}) + \sum\limits_{i} P_{ir2} \tag{10.29}$$

and

$$f(N_o \leqslant n) = P_{ko} \cdot O_k \lambda_1 E + P_{kp} \mu_p + \sum\limits_{i} P_{io} O_i \lambda_1 (1 - E) \tag{10.30}$$

where state k_o is such that for $(k_o + 1)$, $N_o = n$.

10.3.5 Vehicle System Example

The models described in this chapter have been implemented in a computer program [12]. Starting with the component data, the program generates the transition rate matrices for the subsystem and system models and then solves these matrix equations to provide with system-based reliability measures.

The vehicle system data for this example is printed out in Table 10.1A. The system consists of 14 vehicles out of which 2 are on maintenance and two are kept as spares. The assumed failure rate and other data are also listed in Table 10.1A. The probabilities of being in various states (see Figure 10.4) are printed in Table 10.1B. The state description on the right-hand side pertains to *io*, that is, the operating state. Column P contains probabilities of the operating states *io* and the associated columns PR, PS and PP give the probabilities of *ir*2, *ir*1, and *ip*. It can be seen that the most significant probability values are for states $(10, 2, 2)$, $(10, 1, 2)$ and the operating states of $(10, 1, 1)$ and $(10, 1, 0)$. The state probabilities are grouped as a function of the exact number of vehicles in Table 10.1C. The second column, "PROB OF OPTG," gives the probability of being in the operating state with number of vehicles indicated in the first column. The third column "FREQ OF OPTG" can be interpreted in either of the two ways: (a) the number of times the system transits out of the state in a day, or (b) the number of times per day the state is entered by the system. The fourth and fifth columns indicate the probabilities and frequencies of encountering the partial operating states. The probability and frequency of the system being in the retrieval state are given at the bottom. The reliability measures arranged in the cumulative form in Table 10.1D, where the reliability measures for $N_o \leqslant 9$ and downward all are approximately

Table 10.1A *Vehicle system data*

```
MEAN TIME TO VEHICLE FAILURE(PERMANENT) =        5000.0000     HOURS

MEAN TIME TO VEHICLE FAILURE(PARTIAL MODE) =       500.0000     HOURS

MEAN TIME TO VEHICLE RETRIEVAL =     0.5000    HOURS

MEAN TIME TO CLEAR PARTIAL MODE VEHICLE =         0.2500    HOURS

MEAN TIME TO VEHICLE MAINTENANCE =      3.0000     HOURS

MEAN TIME TO VEHICLE REPAIR =      2.5000    HOURS




NUMBER OF OPERATING VEHICLES = 10

NUMBER OF SPARE VEHICLES =   2

NUMBER OF VEHICLES ON MAINTENANCE =   2

VEHICLE EXPOSURE FACTOR = 0.3000
```

Table 10.1B *Vehicle system state probabilities*

THE PROBABILITIES OF BEING IN VARIOUS STATES

STATE NO	PR	PS	P	PP	O	S	M
					STATE DESCRIPTION		
1	0.658554E-03	0.282237E-03	0.940792D 00	0.470396E-02	10	2	2
2	0.357769E-04	0.153330E-04	0.511099D-01	0.255549E-03	10	1	2
3	0.533779E-06	0.228762E-06	0.762541D-03	0.381271E-05	10	0	2
4	0.564417E-08	0.241893E-08	0.895901D-05	0.403156E-07	9	0	2
5	0.437401E-10	0.187458E-10	0.781074D-07	0.312429E-09	8	0	2
6	0.252282E-12	0.108121E-12	0.514861D-09	0.180201E-11	7	0	2
7	0.108501E-14	0.465003E-15	0.258335D-11	0.775005E-14	6	0	2
8	0.344093E-17	0.147468E-17	0.983122D-14	0.245780E-16	5	0	2
9	0.782828E-20	0.335498E-20	0.279582D-16	0.559163E-19	4	0	2
10	0.121067E-22	0.518858E-23	0.576509D-19	0.864764E-22	3	0	2
11	0.114135E-25	0.489148E-26	0.815247D-22	0.815247E-25	2	0	2
12	0.495558E-29	0.212382E-29	0.707940D-25	0.353970E-28	1	0	2
13	0.000000E 00	0.000000E 00	0.000000D 00	0.000000E 00	0	0	2
14	0.265234E-06	0.113672E-06	0.378906D-03	0.189453E-05	10	1	1
15	0.152215E-06	0.652349E-07	0.217450D-03	0.108725E-05	10	0	1
16	0.351460E-08	0.150626E-08	0.557873D-05	0.251043E-07	9	0	1
17	0.390439E-10	0.167331E-10	0.697212D-07	0.278885E-09	8	0	1
18	0.281618E-12	0.120694E-12	0.574731D-09	0.201156E-11	7	0	1
19	0.141748E-14	0.607490E-15	0.337495D-11	0.101248E-13	6	0	1
20	0.506249E-17	0.216964E-17	0.144642D-13	0.361606E-16	5	0	1
21	0.126534E-19	0.542289E-20	0.451908D-16	0.903815E-19	4	0	1
22	0.211348E-22	0.905779E-23	0.100642D-18	0.150963E-21	3	0	1
23	0.212540E-25	0.910887E-26	0.151815D-21	0.151814E-24	2	0	1
24	0.975243E-29	0.417961E-29	0.139321D-24	0.696602E-28	1	0	1
25	0.000000E 00	0.000000E 00	0.000000D 00	0.000000E 00	0	0	1
26	0.529947E-05	0.225406E-06	0.751354D-03	0.375677E-05	10	1	0
27	0.629002E-08	0.269572E-08	0.898575D-05	0.449287E-07	10	0	0
28	0.327709E-09	0.140447E-09	0.520174D-06	0.234078E-08	9	0	0
29	0.787093E-11	0.337325E-11	0.140552D-07	0.562209E-10	8	0	0
30	0.770138E-13	0.330059E-13	0.157171D-09	0.550099E-12	7	0	0
31	0.465970E-15	0.199702E-15	0.110945D-11	0.332836E-14	6	0	0
32	0.189401E-17	0.811721E-18	0.541147D-14	0.135287E-16	5	0	0
33	0.522363E-20	0.223870E-20	0.186558D-16	0.373117E-19	4	0	0
34	0.943872E-23	0.404516E-23	0.449463D-19	0.674194E-22	3	0	0
35	0.996333E-26	0.427000E-26	0.711667D-22	0.711666E-25	2	0	0
36	0.000000E 00	0.000000E 00	0.000000D 00	0.000000E 00	1	0	0
37	0.000000E 00	0.000000E 00	0.000000D 00	0.000000E 00	0	0	0

Table 10.1C *Exact state probabilities and frequencies of the vehicle system*

THE EXACT STATE PROBS AND FREQS

# OF VEH	PROB OF OPTG	FREQ OF OPTG	PROB PARTIAL MODE	FREQ OF PARTIAL MODE
10	0.994319E 00	0.510543E 00	0.497009E-02	0.477130E 00
9	0.153546E-04	0.692341E-03	0.677606E-07	0.650502E-05
8	0.165949E-06	0.903139E-05	0.647535E-09	0.621633E-07
7	0.128561E-08	0.816761E-07	0.436367E-11	0.418912E-09
6	0.732957E-11	0.531376E-09	0.212033E-13	0.203551E-11
5	0.309791E-13	0.251992E-11	0.742673E-16	0.712966E-14
4	0.962607E-16	0.867087E-14	0.183609E-18	0.176265E-16
3	0.214256E-18	0.211488E-16	0.304859E-21	0.292665E-19
2	0.322797E-21	0.346075E-19	0.304506E-24	0.292325E-22
1	0.228385E-24	0.257839E-22	0.105057E-27	0.100855E-25
0	0.630343E-29	0.000000E 00	0.000000E 00	0.000000E 00

RET STATE PROBABILITY= 0.695705E-03

RET STATE FREQUENCY(PER DAY) = 0.333995E-01

Table 10.1D Cumulative state probabilities and frequencies of the vehicle system

OPERATING VEHICLES EQUAL TO OR LESS THAN	PROBABILITY	FREQUENCY PER DAY	CYCLE TIME DAYS	MEAN DURATION DAYS
10	0.100000E 01	0.000000E 00	0.000000E 00	0.000000E 00
9	0.711412E-03	0.338880E-01	0.295089E 02	0.209930E-01
8	0.695990E-03	0.334062E-01	0.299345E 02	0.208341E-01
7	0.695824E-03	0.333996E-01	0.299405E 02	0.208333E-01
6	0.695823E-03	0.333995E-01	0.299405E 02	0.208333E-01
5	0.695823E-03	0.333995E-01	0.299405E 02	0.208333E-01

Table 10.2A   *Vehicle system data*

MEAN TIME TO VEHICLE FAILURE(PERMANENT) =	5000.0000 HOURS
MEAN TIME TO VEHICLE FAILURE(PARTIAL MODE) =	500.0000 HOURS
MEAN TIME TO VEHICLE RETRIEVAL =	0.5000 HOURS
MEAN TIME TO CLEAR PARTIAL MODE VEHICLE =	0.2500 HOURS
MEAN TIME TO VEHICLE MAINTENANCE =	3.0000 HOURS
MEAN TIME TO VEHICLE REPAIR =	2.5000 HOURS

NUMBER OF OPERATING VEHICLES = 10

NUMBER OF SPARE VEHICLES = 0

NUMBER OF VEHICLES ON MAINTENANCE = 0

VEHICLE EXPOSURE FACTOR = 0.3000

equal. This is because of the presence of 2 spares and 2 units on maintenance which also can behave as spares, the probabilities of states with
operating units less than 10 are relatively low and therefore are dominated
by the retrieval state probability.

The results for another example which is the same as the previous one,
except that there are no units spare or on maintenance, are given in Tables
10.2A–D. In Table 10.2D, the reliability measures tend to be almost equal
from $N_o \leqslant 7$ downward as compared with $N_o \leqslant 9$ in Table 10.1D. The
probability and frequency of encountering states with a specified number
of failures, depends upon the failure rate of units, number of spares, and
number of units on maintenance. These relationships will become clearer
in the section on sensitivity studies.

10.4   SYSTEM MODEL FOR TRAINS

When the trains are regarded strictly as single units for operation, spares,
and maintenance, the vehicle system model described in Section 10.3 can
be used. A model, however, is also possible based on a more flexible policy
for train formation. The general procedure is the same as for the vehicle
system:

1. Generate the possibility space by enumerating and describing the possible system states.
2. Develop the transition rate matrix.
3. Solve for probabilities and calculate reliability measures.

Table 10.2B *Vehicle system state probabilities*

THE PROBABILITIES OF BEING IN VARIOUS STATES

					STATE DESCRIPTION		
STATE NO	PR	PS	P	PP	O	S	M
1	0.658709E-03	0.282304E-03	0.9410130 00	0.470506E-02	10	0	0
2	0.326061E-04	0.139740E-04	0.517557D-01	0.232901E-03	9	0	0
3	0.717334E-06	0.307429E-06	0.128095D-02	0.512381E-05	8	0	0
4	0.920579E-08	0.394534E-08	0.187873D-04	0.657556E-07	7	0	0
5	0.759478E-10	0.325490E-10	0.180828D-06	0.542484E-09	6	0	0
6	0.417713E-12	0.179020E-12	0.119346D-08	0.298366E-11	5	0	0
7	0.153161E-14	0.656405E-15	0.547004D-11	0.109401E-13	4	0	0
8	0.361005E-17	0.154716E-17	0.171907D-13	0.257861E-16	3	0	0
9	0.486145E-20	0.208343E-20	0.347247D-16	0.347246E-19	2	0	0
10	0.000000E 00	0.000000E 00	0.000000D 00	0.000000E 00	1	0	0
11	0.000000E 00	0.000000E 00	0.000000D 00	0.000000E 00	0	0	0

Table 10.2C *Exact state probabilities and frequencies of the vehicle system*

THE EXACT STATE PROBS AND FREQS

# OF VEH	PROB OF OPTG	FREQ OF OPTG	PROB PARTIAL MODE	FREQ OF PARTIAL MODE
10	0.941013E 00	0.496855E 00	0.470506E-02	0.451686E 00
9	0.520380E-01	0.521449E 00	0.232901E-03	0.223585E-01
8	0.129493E-02	0.251354E-01	0.512381E-05	0.491886E-03
7	0.190947E-04	0.548019E-03	0.657556E-07	0.631254E-05
6	0.184773E-06	0.700109E-05	0.542484E-09	0.520785E-07
5	0.122601E-08	0.576014E-07	0.298366E-11	0.286431E-09
4	0.564906E-11	0.316230E-09	0.109401E-13	0.105025E-11
3	0.178471E-13	0.115794E-11	0.257861E-16	0.247546E-14
2	0.362718E-16	0.267052E-14	0.347246E-19	0.333357E-17
1	0.208348E-20	0.000000E 00	0.000000E 00	0.000000E 00
0	0.000000E 00	0.000000E 00	0.000000E 00	0.000000E 00

RET STATE PROBABILITY= 0.691772E-03

RET STATE FREQUENCY(PER DAY)= 0.332180E-01

Table 10.2D *Cumulative state probabilities and frequencies of the vehicle system*

OPERATING VEHICLES EQUAL TO	PROBABILITY	FREQUENCY	CYCLE TIME	MEAN DURATION
OR LESS THAN		PER DAY	DAYS	DAYS
10	0.100000E 01	0.000000E 00	0.000000E 00	0.000000E 00
9	0.542823E-01	0.496855E 00	0.201266E 01	0.109252E 00
8	0.201144E-02	0.562123E-01	0.177897E 02	0.357829E-01
7	0.711388E-03	0.337242E-01	0.296523E 02	0.210943E-01
6	0.692228E-03	0.332245E-01	0.300983E 02	0.208349E-01
5	0.692042E-03	0.332180E-01	0.301041E 02	0.208333E-01
4	0.692041E-03	0.332180E-01	0.301042E 02	0.208333E-01
3	0.692041E-03	0.332180E-01	0.301042E 02	0.208333E-01

The state space is generated by a subroutine in the computer program [12] according to the following rules for the train formation.

1. When a vehicle in the train fails, the train is removed from service and a spare train, if available, is put into operation as a replacement. If, however, a complete spare train is not available, the defective vehicle is replaced by a good vehicle and the train is put back into operation. If no good vehicle is available, the defective vehicle is removed and the train put back into operation. There is no additional difficulty in modeling with married pairs as they can be treated as single units.
2. Train-consists of different lengths are allowed.
3. In case of no available spares and when all trains are not full length, a vehicle on which maintenance or repair is completed is attached to the train having the least number of vehicles.

In the rules outlined above, the attempt is to keep the maximum vehicle system capacity in the operating condition. Rules 1–3 represent only one policy and models can be similarly developed for other policies. The results of the solution of the train model are grouped in terms of vehicle system capacity, that is, the output format is the same as for the vehicle system model.

10.5 PASSENGER STATION MODEL

The stations are assumed off-line and basically of two types:

1. Stations having one station lane and an express lane for through traffic. These are termed type *A* stations.
2. Stations having two station lanes and an expressway lane, called type *B* stations.

10.5.1 Model for a Single Station

The state transition diagrams for the type *A* and type *B* stations are shown in Figures 10.6 and 10.7 respectively, using the following notation:

U = normal station operation, that is, both the station lane and the express lane are working
D = station lane down
L = failure of the station lane as well as the express lane, complete failure of type *A* station.
$\bar{L}$ = station lane working but express lane down

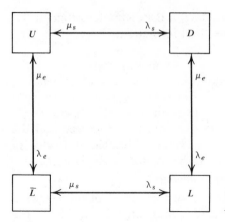

Figure 10.6 State transition diagram of type *A* station.

For the type *B* station:

$2D$ = both the station lanes down
$\bar{L}_1$ = one station lane working and the express lane down
$\bar{L}_2$ = both station lanes working with express lane down
λ_s, μ_s = failure and repair rates of the station lane including the contribution λ_{VS} of the vehicle system failure
λ_e, μ_e = express lane failure and repair rates

The impact of the various station states on the system can be tabulated in Tables 10.3 and 10.4.

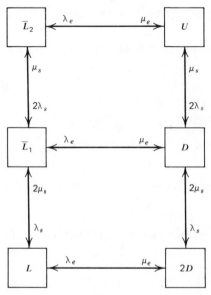

Figure 10.7 State transition diagram of type *B* station.

Table 10.3 *Type A Station*

Station State	Impact on System	Impact on Station
U	Traffic can pass	Passengers can embark/disembark
D	Traffic can pass	Passengers cannot embark/disembark
L	Traffic cannot pass	Passengers cannot embark/disembark
$\bar{L}$	Traffic can pass	Passengers can embark/disembark

10.5.2 Model for System of Stations

A transit system may have $N0$ number of type A stations and $N1$ number of type B stations. The model for the system can be built by combining the models for the individual stations. It is assumed that when at a station both the station lanes and expressway are down and the traffic cannot pass through, no further failure of stations takes place.

The model for the system of stations is built by sequential addition [10]. This is achieved by adding one station at a time, solving for state probabilities, deleting low probability states, and adding the next station. This procedure helps to keep the system states within manageable limits. The process of sequential addition is illustrated in Table 10.5 for three type A stations with the following hypothetical data:

mean up time of the station lane   =   800 hours
mean down time of the station lane =   2 hours
mean up time of the express lane   = 1000 hours
mean down time of the express lane =   2 hours

Table 10.4 *Type B Station*

Station State	Impact on System	Impact on Station
U	Traffic can pass	Passengers can embark/disembark
D	Traffic can pass	Passengers can embark/disembark
$2D$	Traffic can pass	Passengers cannot embark/disembark
L	Traffic cannot pass	Passengers cannot embark/disembark
$\bar{L}_1$	Traffic can pass	Passengers can embark/disembark
$\bar{L}_2$	Traffic can pass	Passengers can embark/disembark

Table 10.5 *Sequential Model Building*

System State	Identical states	No. of stations in state				Probability
		U	D	L	L	
(a) Model of a single station						
1		1	0	0	0	
2		0	1	0	0	
3		0	0	1	0	
4		0	0	0	1	
(b) Addition of a station						
1(1,1)		2	0	0	0	
2(2,1)		1	1	0	0	
3(3,1)		1	0	1	0	
4(4,1)		1	0	0	1	
5(1,2)		1	1	0	0	
6(2,2)		0	2	0	0	
7(3,2)		0	1	1	0	
8(4,2)		0	1	0	1	
9(1,3)		1	0	1	0	
10(2,3)		0	1	1	0	
11(4,3)		0	0	1	1	
12(1,4)		1	0	0	1	
13(2,4)		0	1	0	1	
14(3,4)		0	0	1	1	
15(4,4)		0	0	0	2	
(c) Merging of identical states						
1	1	2	0	0	0	0.991051
2	2,5	1	1	0	0	0.495525×10^{-2}
3	3,9	1	0	1	0	0.992536×10^{-5}
4	4,12	1	0	0	1	0.396420×10^{-2}
5	6	0	2	0	0	0.618994×10^{-5}
6	7,10	0	1	1	0	0.165058×10^{-7}
7	8,13	0	1	0	1	0.990308×10^{-5}
8	11,14	0	0	1	1	0.132036×10^{-7}
9	15	0	0	0	2	0.396090×10^{-5}
(d) Truncation of states with probabilities less than 10^{-5}						
1		2	0	0	0	0.991051
2		1	1	0	0	0.495525×10^{-2}
3		1	0	1	0	0.992536×10^{-5}
4		1	0	0	1	0.396420×10^{-2}
5		0	2	0	0	0.618994×10^{-5}
6		0	1	0	1	0.990308×10^{-5}
7		0	0	0	2	0.396090×10^{-5}
(e) Addition of the third station						
1(1,1)		3	0	0	0	
2(2,1)		2	1	0	0	
3(3,1)		2	0	1	0	
4(4,1)		2	0	0	1	

TABLE 10.5 *(Continued)*

System State	Identical states	No. of stations in state				Probability
		U	D	L	$\bar{L}$	
5(5,1)		1	2	0	0	
6(6,1)		1	1	0	1	
7(7,1)		1	0	0	2	
8(1,2)		2	1	0	0	
9(2,2)		1	2	0	0	
10(3,2)		1	1	1	0	
11(4,2)		1	1	0	1	
12(5,2)		0	3	0	0	
13(6,2)		0	2	0	1	
14(7,2)		0	1	0	2	
15(1,3)		2	0	1	0	
16(2,3)		1	1	1	0	
17(4,3)		1	0	1	1	(3,3) not possible
18(5,3)		0	2	1	0	
19(6,3)		0	1	1	1	
20(7,3)		0	0	1	2	
21(1,4)		2	0	0	1	
22(2,4)		1	1	0	1	
23(3,4)		1	0	1	1	
24(4,4)		1	0	0	2	
25(5,4)		0	2	0	1	
26(6,4)		0	1	0	2	
27(7,4)		0	0	0	3	
(f) *Merging of identical states*						
1	1	3	0	0	0	0.986606
2	2,8	2	1	0	0	0.739954×10^{-2}
3	3,15	2	0	1	0	0.148435×10^{-4}
4	4,21	2	0	0	1	0.591964×10^{-2}
5	5,9	1	2	0	0	0.184865×10^{-4}
6	6,11,22	1	1	0	1	0.295760×10^{-4}
7	7,24	1	0	0	2	0.118294×10^{-4}
8	10,16	1	1	1	0	0.493508×10^{-7}
9	12	0	3	0	0	0.153901×10^{-7}
10	13,25	0	2	0	1	0.369310×10^{-7}
11	14,26	0	1	0	2	0.295407×10^{-7}
12	17,23	1	0	1	1	0.394773×10^{-7}
13	18	0	2	1	0	0.461670×10^{-10}
14	19	0	1	1	1	0.738569×10^{-10}
15	20	0	0	1	2	0.295387×10^{-10}
16	27	0	0	0	3	0.787643×10^{-8}

Table 10.6 *Model of Three Stations Without Truncation*

System State	State No. as in Table 10.1	No. of stations in U	D	L	L	Probability
1	1	3	0	0	0	0.986606
2	2	2	1	0	0	0.739954×10^{-2}
3	3	2	0	1	0	0.148435×10^{-4}
4	4	2	0	0	1	0.591964×10^{-2}
5	5	1	2	0	0	0.184865×10^{-4}
6	8	1	1	1	0	0.493508×10^{-7}
7	6	1	1	0	1	0.295760×10^{-4}
8	12	1	0	1	1	0.394773×10^{-7}
9	7	1	0	0	2	0.118294×10^{-4}
10	9	0	3	0	0	0.153901×10^{-7}
11	13	0	2	1	0	0.461824×10^{-10}
12	10	0	2	0	1	0.369310×10^{-7}
13	14	0	1	1	1	0.738538×10^{-10}
14	11	0	1	0	2	0.295407×10^{-7}
15	Deleted	0	1	2	0	0.123121×10^{-12}
16	Deleted	0	0	2	1	0.984465×10^{-13}
17	15	0	0	1	2	0.295264×10^{-10}
18	16	0	0	0	3	0.787643×10^{-8}

The model for a single station is represented in Table 10.5*a*. The addition of one more station is shown in Table 10.5*b*. For each state of the station being added, there is a set of system states of Table 10.5*a* except the state (3,3). The state (3,3) representing two stations completely down (state L) is assumed not possible since exposure to failure is assumed zero as soon as one station is completely down. The system states in Table 10.5*b* are numbered in the serial order and the numbers in parenthesis indicate the combination, the first number denoting the system state before addition and the second indicating the state of the station being added. The identical states can now be grouped together as shown in Table 10.5*c*. The states with probabilities less than 10^{-5} (an arbitrary reference) are deleted and the remaining states are shown in Table 10.5*d*, where the state numbers are serial and have no relationship to state numbers in Table 10.5*c*. Tables 10.5*e* and 10.5*f* show the addition to the third station. If a fourth station is now to be added, then the states with probabilities less than 10^{-5} can again be deleted and the procedure repeated. The exact results, without any truncation, are shown in Table 10.6 and are almost identical to Table 10.5*f*. In general, the effect on the results depends on the reference probability value employed for truncation.

10.6 MODELS FOR OTHER SUBSYSTEMS

10.6.1 Power Substations Model

A single power station is assumed to have only two states, up (i.e., the substation is working) and down (i.e., the substation is failed). The system of substations is, however, assumed as an m/n configuration; that is, the system is good if m out of n substations are working. In other words the assumption means that so long as m out of n substations are working, the power supply is adequate to keep the system running. When, however, one more substation fails, the system either completely fails or goes into severe degradation. The state transition diagram for the substations model is shown in Figure 10.8, where λ_{ss} and μ_{ss} are the substation failure and repair rates respectively.

10.6.2 Guideway Model

Guideway consists of the structure, power rails, and any other equipment whose failure would incapacitate the guideway. As an example in magnetic

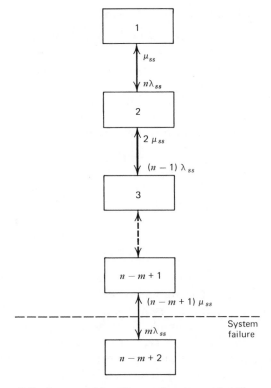

Figure 10.8 State transition diagram for power substations model.

levitation and linear induction propulsion, the guideway would include suspension armature rail and LIM rail. The guideway is assumed to be a two-state system: Up state when the guideway is in satisfactory condition and down state when a failure of guideway interrupts the flow of traffic.

10.6.3   Command and Control System Model

Command and control includes all equipment associated with the control of vehicle movement, except the vehicle-borne equipment, which is regarded as a part of the vehicle. Like the guideway, the command and control is also assumed a two-state system.

10.7   TOTAL SYSTEM MODEL

The models for various subsystems have been described in Sections 10.3–10.6. These models can be combined to give the reliability measures of the entire transit system.

10.7.1   System-Based Reliability Measures

The various subsystem models can be combined by the sequential model building described in Section 10.5.2. The subsystem models, for this purpose, are reduced to the following equivalent forms:

1. *Vehicle/train system model.* The vehicle or train system model is represented by a two-state equivalent such that State $1=(N_0>n)$ and State $2=(N_0 \leqslant n)$. The equivalent transition rates are

$$\lambda_{12} = \frac{f(N_0 \leqslant n)}{P(N_0 > n)}$$

and

$$\lambda_{21} = \frac{f(N_0 \leqslant n)}{P(N_0 \leqslant n)}$$

2. *Station system model.* The station system model is reduced to a multistate model whose states represent the possible combination of stations in the working state, for example, for a system of 3 type A and 1 type B stations, there could be $(3,1)$, $(2,1)$, $(0,0)$ states, where the numbers in the parenthesis indicate the number of type A and type B stations in the up state. The equivalent transition rates between the various states can be determined using (10.3).

3. *Substation Model.*

State 1 = (number of up stations ≥ m)
State 2 = (number of up stations < m)
The equivalent transition rates are

$$\lambda_{12} = P(N_{ss} = m) \cdot m \frac{\lambda_{ss}}{P(N_{ss} \geq m)}$$

$$\lambda_{21} = \frac{P(N_{ss} = m-1)(n-m+1)\mu_{ss}}{P(N_{ss} < m)}$$

where N_{ss} is the number of working substations.

The station system model, the substation system model, the command and control model and the guideway model are combined together and the resulting states are grouped as $(N_A = i, N_B = j)$ where N_A and N_B indicate the number of type A and type B stations up, respectively. The state $N_A = 0, N_B = 0$ includes the condition of having the substation system, guideway or command and control down. Therefore, $(N_A = 0, N_B = 0)$ does not necessarily mean that all the stations are failed. It really means that the stations are not available for embarcation because the system is not moving. This combined model is then combined with the vehicle system model to give the measure for the entire system.

The printout of the reliability measures of the system is shown in Tables 10.1I–10.1L. The data for the vehicle system are shown in Table 10.1A and for other subsystems in Tables 10.1E–10.1H.

10.7.2 Including Demand [13]

The calculation of probabilities, frequencies, and the mean duration of the various deficient states of the system is illustrated in Section 10.7.1. These system deficiencies can be further related to the delays that they cause to the vehicles or passengers and then the probabilities of these delays can be

Table 10.1E *Passenger stations data*

NUMBER OF TYPE A STATIONS = 3	
NUMBER OF TYPE B STATIONS = 1	
MEAN TIME TO FAILURE OF THE STATION = 3260.88 HOURS	
MEAN TIME TO REPAIR OF THE STATION = 1.97 HOURS	
MEAN TIME TO FAILURE OF THE EXPRESSWAY = 466666.80 HOURS	
MEAN TIME TO REPAIR OF THE EXPRESSWAY = 10.00 HOURS	

Table 10.1F *Power substations data and results*

```
NUMBER OF SUB STATIONS =   4

MINIMUM NUMBER OF SUB STATIONS REQUIRED FOR OPERATION =   2

MEAN TIME TO SUB STATION FAILURE = 20000.00    HOURS

MEAN TIME TO SUB STATION REPAIR =     0.50    HOURS



SUB STATION SYSTEM AVAILABILITY =   0.100000D 01

FREQUENCY OF ENCOUNTERING THE DOWN STATE =  0.899910D-11    PER DAY

CYCLE TIME TO ENCOUNTER DOWN STATE =  0.111122D 12    DAYS

MEAN DURATION OF DOWN STATE =  0.692571D-02    DAYS
```

Table 10.1G *Command and control system data and results*

```
MEAN TIME TO COMMAND AND CONTROL FAILURE =  1666.67    HOURS

MEAN TIME TO COMMAND AND CONTROL REPAIR =     2.58    HOURS



COMMAND AND CONTROL AVAILABILITY =   0.998454E 00

FREQUENCY OF ENCOUNTERING DOWN STATE =  0.143777E-01    PER DAY

CYCLE TIME TO ENCOUNTER DOWN STATE =  0.695521E 02    DAYS

MEAN DURATION OF DOWN STATE =  0.107500E 00    DAYS
```

Table 10.1H *Guideway data and results*

```
MEAN TIME TO GUIDE WAY FAILURE = 50000.00    HOURS

MEAN TIME TO GUIDE WAY REPAIR =     10.00    HOURS



GUIDE WAY AVAILABILITY =   0.999800E 00

FREQUENCY OF ENCOUNTERING DOWN STATE =  0.479904E-03    PER DAY

CYCLE TIME TO ENCOUNTER OF DOWN STATE =  0.208375E 04    DAYS

MEAN DURATION OF DOWN STATE =  0.416447E 00    DAYS
```

Table 10.1I *System reliability measures*

# OF TYPE A STATIONS UP	#OF TYPE B STATIONS UP	# OF VEHICLES GREATER THAN	PROBABILITY	FREQUENCY PER DAY	CYCLE TIME DAYS	MEAN DURATION
3	1	9	0.995740D 00	0.705775E-01	0.141688E 02	14.108
2	1	9	0.180329D-02	0.220835E-01	0.452827E 02	0.082
1	1	9	0.108763D-05	0.265535E-04	0.376597E 05	0.041
0	0	9	0.174413D-02	0.149114E-01	0.670628E 02	0.117

# OF TYPE A STATIONS UP	# OF TYPE B STATIONS UP	# OF VEHICLES EQUAL OR LESS THAN	PROBABILITY	FREQUENCY PER DAY	CYCLE TIME DAYS	MEAN DURATION
3	1	9	0.708836D-03	0.337939E-01	0.295911E 02	0.021
2	1	9	0.128380D-05	0.768316E-04	0.130155E 05	0.017
1	1	9	0.774302D-09	0.557615E-07	0.179335E 08	0.014
0	0	9	0.124168D-05	0.697208E-04	0.143429E 05	0.018

Table 10.1J *System reliability measures*

# OF TYPE A STATIONS	#OF TYPE B STATIONS	# OF VEHICLES GREATER	PROBABILITY	FREQUENCY	CYCLE TIME	MEAN
UP	UP	THAN		PER DAY	DAYS	DURATION
3	1	8	0.995755D 00	0.700980E-01	0.142657E 02	14.205
2	1	8	0.180332D-02	0.220830E-01	0.452838E 02	0.082
1	1	8	0.108764D-05	0.265534E-04	0.376599E 05	0.041
0	0	8	0.174416D-02	0.149108E-01	0.670655E 02	0.117

# OF TYPE A STATIONS	# OF TYPE B STATIONS	# OF VEHICLES EQUAL	PROBABILITY	FREQUENCY	CYCLE TIME	MEAN
UP	UP	OR LESS THAN		PER DAY	DAYS	DURATION
3	1	8	0.693519D-03	0.333132E-01	0.301181E 02	0.021
2	1	8	0.125597D-05	0.756222E-04	0.132236E 05	0.017
1	1	8	0.757517D-09	0.548279E-07	0.182389E 08	0.014
0	0	8	0.121476D-05	0.686506E-04	0.145665E 05	0.018

Table 10.1K *System reliability measures*

# OF TYPE A STATIONS UP	#OF TYPE B STATIONS UP	# OF VEHICLES GREATER THAN	PROBABILITY	FREQUENCY PER DAY	CYCLE TIME DAYS	MEAN DURATION
3	1	7	0.995756D 00	0.700914E-01	0.142671E 02	14.207
2	1	7	0.180332D-02	0.220830E-01	0.452838E 02	0.082
1	1	7	0.108764D-05	0.265534E-04	0.376599E 05	0.041
0	0	7	0.174416D-02	0.149108E-01	0.670656E 02	0.117

# OF TYPE A STATIONS UP	# OF TYPE B STATIONS UP	# OF VEHICLES EQUAL OR LESS THAN	PROBABILITY	FREQUENCY PER DAY	CYCLE TIME DAYS	MEAN DURATION
3	1	7	0.693353D-03	0.333066E-01	0.300240E 02	0.021
2	1	7	0.125567D-05	0.756066E-04	0.132264E 05	0.017
1	1	7	0.757336D-09	0.548162E-07	0.182428E 08	0.014
0	0	7	0.121447D-05	0.686366E-04	0.145695E 05	0.018

Table 10.1L *System reliability measures*

# OF TYPE A STATIONS UP	#OF TYPE B STATIONS UP	# OF VEHICLES GREATER THAN	PROBABILITY	FREQUENCY PER DAY	CYCLE TIME DAYS	MEAN DURATION
3	1	6	0.9957560 00	0.700913E-01	0.142671E 02	14.207
2	1	6	0.180332D-02	0.220830E-01	0.452838E 02	0.082
1	1	6	0.108764D-05	0.265534E-04	0.376599E 05	0.041
0	0	6	0.174416D-02	0.149108E-01	0.670656E 02	0.117

# OF TYPE A STATIONS UP	# OF TYPE B STATIONS UP	# OF VEHICLES EQUAL OR LESS THAN	PROBABILITY	FREQUENCY PER DAY	CYCLE TIME DAYS	MEAN DURATION
3	1	6	0.693352D-03	0.333066E-01	0.300241E 02	0.021
2	1	6	0.125566D-05	0.756065E-04	0.132264E 05	0.017
1	1	6	0.757335D-09	0.548162E-07	0.182428E 08	0.014
0	0	6	0.121447D-05	0.686365E-04	0.145695E 05	0.018

calculated. The extent to which such an analysis can be carried out depends upon the information available on the flow of passengers. The calculation of delays can be illustrated by an example. Consider a system such that only vehicle fleet need be considered and the other subsystems can be ignored. The delay could be caused by the following types of system deficiencies,

1. A vehicle could fail in the retrieval mode blocking the flow of vehicles. This condition will last till the failed vehicle has been removed and the system put back into normal operation. Let the probability of being in this state be denoted by P_{RET}.
2. A vehicle could fail in a partial mode, degrading the operation of the system until the vehicle finally clears the system. Let the probability of being in this state be denoted by P_{PR}.
3. The delay could also be caused because the number of vehicles available for service is less than the required number. This condition can result when spare vehicles are not available to replace the failed ones.

These probabilities can be calculated using the models and methods described in this report. These probabilities can be then weighted with the probabilities of delay caused by these conditions. This will yield the probability of delay caused by system deficiencies. The calculation of probabilities of delay by the deficient conditions is not covered in this chapter.

10.8 SYSTEM STUDIES

Some vehicle system sensitivity studies using these models are reported here. The system is assumed to consist of 50 operating vehicles and the relevant data is listed on appropriate figures. The vehicle system state with number of vehicles less than or equal to 49 is considered as the reference state.

10.8.1 Sensitivity of Reliability Indices to the Vehicle MTBF (Retrieval Mode)

The effect of variation in vehicle MTBF on the probability, mean time to encounter and the mean duration of the system state with vehicles $\leqslant 49$ is shown in Figures 10.9–10.11. The reliability indices are plotted for three cases, $(s=2, m=2)$, $(s=0, m=2)$ and $(s=0, m=0)$. The probability of $N_o \leqslant 49$ is the lowest and the most sensitive in the case of $s=2$, $m=2$. This is because when there are no spare vehicles, the vehicle failure rate in partial mode begins to be effective and since it is 500 hours as compared with 5000 hours for the retrieval mode, the partial failure mode dominates

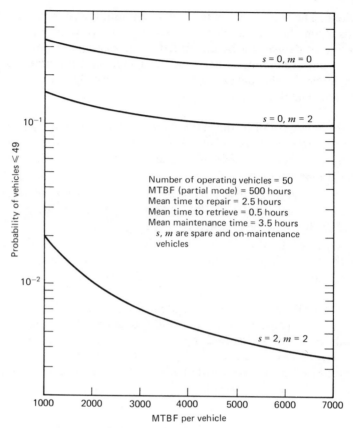

Figure 10.9   System failure probability vs vehicle MTBF.

for $s=0$. The mean time to encounter $N_o \leqslant 49$ is again highest for $s=2$, $m=2$ and shows the most sensitivity to variation in retrieval mode vehicle MTBF. This is because for $s=0$, the partial mode of failure dominates. The mean duration of $N_o \leqslant 49$ is insensitive to the variation in vehicle MTBF since with spare vehicles available to replace the failed ones, this index is more or less determined by the mean time to retrieve.

10.8.2   Sensitivity of Reliability Indices to the Mean Time to Repair a Vehicle

The effect of variation in vehicle MTTR on the reliability indices of the system is shown in Figures 10.12–10.14. The probability and the mean duration of $N_o \leqslant 49$ decrease with the increase in the MTTR and the mean time to encounter $N_o \leqslant 49$ correspondingly increases. For $s=0$, the mean time to encounter $N_o \leqslant 49$ is relatively insensitive to the MTTR. This is because in this case, the system behaves more or less like a series system

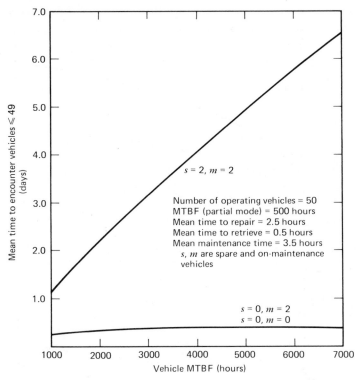

Figure 10.10 System mean time to vehicles ≤ 49 vs vehicle MTBF.

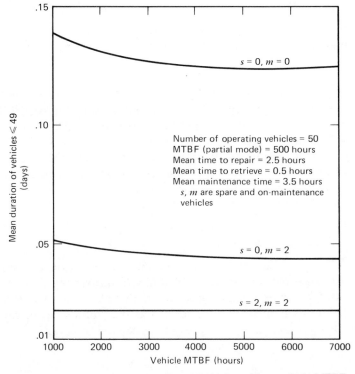

Figure 10.11 System mean duration of (vehicles ≤ 49) vs vehicle MTBF.

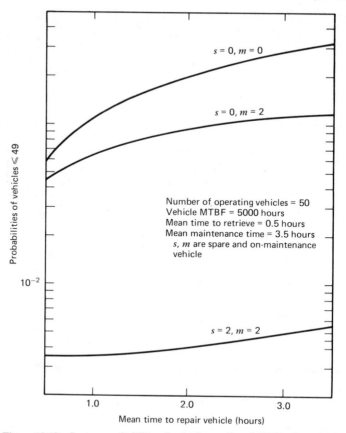

Figure 10.12 System probability of vehicles < 49 vs MTTR of a vehicle.

and the mean time to encounter $N_o \leq 49$ is controlled by the vehicle MTBF (both partial and retrieval mode). The sensitivity of the indices to the vehicle MTTR also depends on the ratio of the spare vehicles to the operating vehicles. The higher the ratio, the less sensitive the indices are to MTTR [6] because with the spare vehicles available to replace the failed ones, the retrieval time dominates the time to repair.

10.8.3 Effect of Spare Vehicles

Figure 10.15 shows the effect of the number of spares on the reliability indices. As expected the system reliability improves by having spare vehicles. After a certain number of spares, the effect is, however, incremental small, and this number may be termed as the "infinite spare capacity" for the system, as there is little improvement beyond this point.

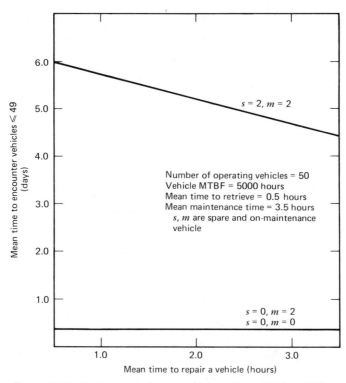

Figure 10.13 System mean time to vehicles ≤ 49 vs vehicle MTTR.

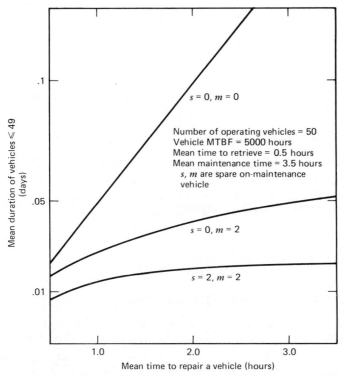

Figure 10.14 System mean duration of vehicles < 49 vs vehicle MTTR.

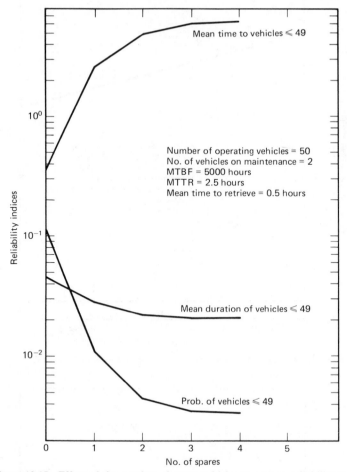

Figure 10.15 Effect of the number of spare vehicles on system reliability indices.

10.8.4 Effect of Entraining Vehicles

A study on the effect of entraining is given in reference 6. The process of entraining vehicles modifies the modes of failure by converting some retrieval mode failures into partial modes and introducing some additional elements for failure, for example, couplers and trainlines.

10.9 CONCLUDING REMARKS

10.9.1 Effect of Peak and Off Peak Periods

The state space reliability models for track bound transit systems have been described and the calculation of the system and demand based reliability measures discussed. This work was originally carried out for application to a single loop configuration. The methodology can be applied to more complicated networks.

Since these models were developed for application to a demonstration project, the effect of peak and offpeak periods was ignored. Therefore these models can find direct application for demonstration, airport or downtown people mover systems where the ratio of peak to off peak periods is closer to unity.

There are two ways [13] of including the effect of peak and off peak periods:

1. If it can be assumed that all the failed vehicles over the previous day have been repaired by the following morning so that the initial probability vector every morning is the same, the models can be solved in a time-specific manner over the peak and off peak periods. This will involve the solution of differential equations, instead of the linear equations for the state probabilities.
2. The models can be modified to include the effect of peak and off peak periods. The approach outlined previously in (1) is, however, simpler to implement.

10.9.2 Simulation Versus Analytical Method

As the systems become more complex, the analytical techniques become more difficult to apply. Simulation using Monte-Carlo techniques can be used to perform the reliability analysis of track-bound systems. The simulation method is conceptually easy to apply but could be quite expensive for sensitivity studies. Sometimes it may be possible to apply a hybrid approach, that is, part solution by analytical methods and part by simulation. For example, the system base probabilities may be calculated by analytical models and the probabilities of delays by specific system deficiencies calculated by simulation and the two results combined to yield demand based measures.

10.9.3 Failure Data

Subsystem failure rate is an important input parameter for transit system reliability modeling and calculation. For transit systems using newer technologies, these figures are usually synthesized from part failure rates available from handbooks or other data collection and exchange programs. Although such information on conventional transit systems could be available from field experience, no collective effort at national or international level has been visible in this regard. A data collection and analysis activity to fulfill this need has been carried out and the results are reported in references 8 and 14.

10.9.4   Further Work

The simulation techniques are conceptually simpler for calculating the reliability measures but consume considerable computer time, especially when performing sensitivity studies. The analytical methods become quite complicated when applied to complex system configurations but are very suitable for sensitivity analysis. There is a need for further development of the analytical and simulation methods for application to more complex network configurations, taking into account the peak and off peak periods, and including demand in deriving suitable measures of reliability.

REFERENCES

1. Billinton, R. and C. Singh, "Frequency and Duration Method of Generating Capacity Reliability Evaluation," Transactions of the Engineering Institute of Canada, vol. 15, March 1972, pp. I–V.

2. Heiman, D., "Availability—Concepts and Definitions," *Proceedings 1976 Annual Reliability and Maintainability Symposium*, IEEE, New York, 1976.

3. Welker, E. L. and H. H. Buchanan, "Safety—Availability Study Methods Applied to BART," *Proceedings 1975 Annual Reliability and Maintainability Symposium*, IEEE, New York, 1975.

4. Welker, E. L., "Reliability and Availability Assessment Criteria, Data Inputs and Analysis Methods for Mass Transit Systems," *Proceedings 1976 Annual Reliability and Maintainability Symposium*, IEEE, New York, 1976.

5. Westmoreland, M. E., "Reliability Improvement of Land Vehicles," *Proceedings 1973 Annual Reliability and Maintainability Symposium*, IEEE, New York, 1973.

6. Singh, C., R. Billinton, J. H. Parker, and R. Puccini, "Transit System Reliability," presented at 2nd Intersociety Transportation Conference, Denver, 1973.

7. Singh, C., "Some Reliability Data on Conventional Transit Vehicles," internal report Transit Systems R & D Branch, Ministry of Transportation and Communications, Ontario, Canada, Nov. 1974.

8. Singh, C. and M. D. Kankam, "Reliability Data and Analysis for Transit Vehicles," Research Report No. RR217, Transit Systems R & D Branch, Ministry of Transportation and Communications, Ontario, Canada, Jan. 1977.

9. Singh, C., "Reliability Models for Track Bound Transit Systems," *Proceedings 1977 Annual Reliability and Maintainability Symposium*, IEEE, New York, 1977.

10. Singh, C. and R. Billinton, *System Reliability Modelling and Evaluation*, Hutchinson Educational, London, 1977.

11. Singh, C., "Operational Reliability Analysis," internal report Transit Systems R & D Branch, Ministry of Transportation and Communications, Ontario, Canada, Feb. 1975.

12. Singh, C., "Computer Program for Reliability Analysis of Track Bound Transit Systems," information available from the author.

13. Singh, C., "Reliability Analysis of Track Bound Transit Systems," Research Report, Research and Development Division, Ministry of Transportation and Communications, Ontario, Canada, May 1977.

14. Singh, C. and Kankam, M. D., "Failure Data Analysis for Transit Vehicles," *Proceedings of the 1979 Annual Reliability and Maintainability Symposium*, IEEE, New York, 1979.

11

Computer System Reliability Modeling

11.1 INTRODUCTION

The scope of computer applications has steadily expanded since the early 1960s, encompassing numerous areas of critical importance. These applications include existing and proposed uses in real time control of communication and transportation systems, automated plant operations and space explorations. These systems generally demand high reliability since system failures can be very costly and hazardous. Very significant progress has been made in the field of computer reliability both in the simplex sense and in the use of redundancy to achieve this objective. This chapter reviews different basic approaches to computer reliability and classification of computer faults and describes modeling of permanent and transient faults.

The term reliability in the context of a computer or information processing system pertains to the correct execution of a program. The modeling and assurance of reliability in digital systems becomes somewhat different from many other systems because of the lack of ability of digital systems to tolerate temporary disturbances. A transient disturbance in many systems would degrade the system performance temporarily, after which the system is restored to normal operation. In digital systems, a transient disturbance could become fatal to a computation, if at a critical moment contents of a register are changed. This subtle difference between the analog and digital systems [2] has not generally been perceived by many reliability analysts.

11.2 CAUSES OF COMPUTER FAILURES [15-20]

For designing reliable computer-based systems, it is essential to identify causes of failures in computers. This section provides a list of the various sources of failure. It is to be noted that failures due to hardware components are only one of such causes. Therefore, reliability predictions based on these failures could be optimistic. A discussion of some important sources of errors is provided in this section. Readers interested in this topic will find references 15–20 of great value.

Memory and Processor Failures.   Modern computers generally use a parity bit to detect failures in memory. Memory failures can be serious as they can cause the entire system to shut down as the operating system can not effectively deal with this problem. Power surges and failures can cause memory problems.

Processor errors are rare but generally catastrophic. Reference to index register *n* may, for example, be suddenly changed due to a dropped bit and this will undoubtedly cause the system to go berserk.

Peripheral Device Failures.   Failure or degradation of peripheral device hardware can sometimes cause serious problems, although generally they do not result in system shut down.

Intermodule Communication Failures.   It is generally accepted that communication failures do occur and will continue to occur. Various error-detecting and -correcting codes are employed but nevertheless some communication errors do finally result in outages of terminals and lines.

Human Errors.   The two important sources of human errors are the operator errors and errors in the software. The matter of software reliability has been covered in Chapter 5. The operators occasionally cause a system crash by starting or shutting down system incorrectly or by incorrectly responding to a particular situation.

Environmental Failures.   The environmental failures can result from electromagnetic interference due to inadequate shielding and by the failure of air conditioning equipment.

Power Failures.   A strong power surge could seriously degrade the life expectancy of electronic components and cause many lingering problems. The computer systems are sensitive even to transient dips and surges and must be provided with proper protection.

The various possible sources of computer faults have been listed. When a computer failure does, however, occur it is not easy to classify the exact cause of failure. Many of the faults remain unexplained.

11.3   CLASSIFICATION OF FAULTS

Despite the numerous possible locations of faults described in the previous section, they basically originate either from permanent failures of hardware components, temporary malfunctions of these components, or external interference with the computer system operation. For the purposes of reliability modeling and evaluation, a useful way of classifying faults is on the basis of their duration. The faults may be classified as transient or permanent.

Permanent faults are often caused by the catastrophic failures of the components. The failure of the component in this case is irreversible and permanent and needs repair or replacement. These faults are characterized by long duration and have a failure rate depending on the environment. A component, for example, will have generally a different failure rate in power-on and power-off situations [12].

The transient faults, on the other hand, are caused by temporary malfunction of the components or by the external interference such as electrical noise, power dips, and glitches. These faults are of limited duration and although they require restoration do not involve repair or replacement. These faults are characterized by arrival modes and duration of transients [3].

11.4 BASIC APPROACHES TO COMPUTER RELIABILITY

Like other physical systems the reliability of computer systems may be enhanced either in a simplex manner or through the use of some form of redundancy. The simplex approach would involve use of high reliability components, conservative designs and adequate attention to fabrication and assembly. In this type of approach, which has also been called the fault intolerant approach, correct program execution implies fault-free hardware operation. There has been a tremendous improvement in the reliability of computer components, making simplex configurations more reliable. The improvement in component reliability, however, has been accompanied by a corresponding growth in the complexity of computer systems requiring an ever-increasing number of components. It has been recognized for a long time [13] that simplex configurations are not likely to provide the ultrareliable computer systems required for space explorations and real-time control of ground or airborne systems.

An alternative and complementary approach to reliable computer systems is that of fault-tolerance. Here the faults are expected to occur but disruptive effect of faults is avoided or minimized by providing some form of redundancy. The computer system can tolerate a predetermined number of faults and execute the programs correctly in their presence. The fault tolerance in computer systems is provided by protective redundancy, which may be implemented in three different ways.

1. Hardware redundancy, that is, additional hardware in terms of redundant logic and/or replication of entire computers.
2. Software redundancy, that is, additional programs for either masking faults or providing recovery.
3. Time redundancy in terms of repetition of machine operations.

Functionally redundancy may be either static or dynamic.

11.4.1 Static Redundancy

Fault Masking. The effect of faults may be masked by providing additional hardware such that the output of the module remains unaffected even though the fault has occurred in one of the redundant units. The effect of a faulty component is instantaneously and automatically masked by permanently connected and concurrently operating units [5, 7]. The reason for naming it static redundancy is that fault masking is autonomous and is without intervention through input-output terminals.

Majority voting redundancy which is the most prominent form of fault masking was proposed by John von Neuman [11] who developed and analyzed triplication of units with majority voting. This type of redundancy has been made economically feasible by integrated circuit technology. One of the interesting illustrations of this approach is SATURN V launch vehicle computer. The SATURN V computer employs a triple modular redundancy with voting elements in the central processor and duplication in the main memory [8].

Application of Coding Theory. Error detecting and correcting codes developed in connection with communication theory and special codes developed for high-speed encoding and decoding have been used for implementing fault tolerance in data transmission and storage functions. Reference 7 states that cost of implementing such schemes is less than 1.5 times the nonredundant configuration.

11.4.2 Dynamic Redundancy

In dynamic redundancy, the fault effect is allowed to appear at the terminals but means are provided for detection, diagnosis, and recovery. If human intervention is eliminated, dynamic redundancy results in computer system self-repair. The first operational computer with full self-repair is probably the JPL-STAR computer [1]. This type of redundancy has also been termed standby redundancy. Error correction is achieved by recomputation, possibly retracing the steps in the program to a roll back point. Reference 5 provides a qualitative and quantitative comparison of the static and dynamic redundancy techniques.

11.4.3 Hybrid Redundancy

In such a scheme, at any moment, three or more elements are connected to a majority element. When a module, however, fails its disagreement with its companions is detected and it is replaced by a spare unit.

11.5 MODELING PERMANENT FAULTS

Modeling of alternative configurations provides information for the judicious choice for a particular application. A number of redundant configurations have been proposed for computer architecture. Models for more commonly known configurations are discussed. The reliability measures depend on the nature of the application. For space-borne equipment, where the equipment is not accessible to repair, the probability of successful operation during the mission time is an appropriate measure. On the other hand, for applications like automated transit or plant control, a measure like mean up time, mean down time, and failure frequency also provide useful information. The failure rates and repair rates where applicable are assumed constant.

11.5.1 Simplex System

The reliability of a simplex system is well-known,

$$R = e^{-\lambda T} \tag{11.1}$$

where λ = failure rate of the system
T = mission time

Equation 11.1 gives the probability that the system will not have failed by mission time T. For repairable system

$$A = \frac{\mu}{\lambda + \mu} \tag{11.2}$$

$$\bar{A} = \frac{\lambda}{\lambda + \mu} \tag{11.3}$$

$$f_f = \frac{\lambda \mu}{\lambda + \mu} \tag{11.4}$$

where A = system availability, that is, long run average proportion of time in success state
$\bar{A}$ = unavailability, that is, long run proportion of time spent in failed state
f_f = frequency of failure, that is, average number of failures per unit time of system operation

11.5.2 Triple-Modular Redundant System

Triple modular redundancy (TMR) is probably the most tossed around term in computer redundancy reliability. The basic TMR system is shown in Figure 11.1, which consists of three identical units representing a single logical variable, the value of the variable being determined on the basis of majority voting. The function of the voter is illustrated in Table 11.1. If there is an independent fault in one of the units, it is evidently masked and output remains correct.

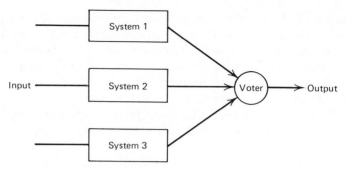

Figure 11.1 Triple-modular redundant system with voting.

The reliability of the TMR configuration is given by the following equation:

$$R_{\text{TMR}} = R^3 + 3(1-R)R^2$$

$$= R^2(3-2R) \tag{11.5}$$

In deriving (11.5) the voter is assumed perfectly reliable and the effect of compensating failures is ignored. For a nonperfect voter, (11.5) becomes

$$R_{\text{TMR}} = R^2(3-2R)R_V \tag{11.6}$$

where R = reliability of each unit
R_V = voter reliability

In reality some of the failures can be compensatory. For example if one of the units is stuck at 0 and the other is stuck at 1, then their votes cancel and if, the third unit is operating correctly, the output will be correct.

Table 11.1 *TMR Voting*

	Unit Output		
System 1	System 2	System 3	Voter Output
0	0	0	0
0	0	1	0
0	1	0	0
1	0	0	0
0	1	1	1
1	0	1	1
1	1	0	1
1	1	1	1

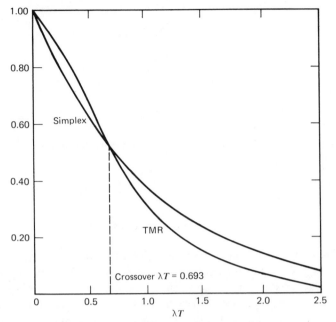

Figure 11.2 Reliability of TMR vs SIMPLEX.

Equations 11.5 and 11.6 therefore provide a pessimistic or conservative estimate. This topic is further discussed in reference 9. The behavior of TMR with respect to the simplex system is shown in Figure 11.2 over the normalized mission time. It is to be noted that whereas reliability of redundant configuration is better than the simplex system until the cross-over point, the reliability of TMR is worse after this point. This type of relative behavior has led to the suggestion [4] that MTTF may be a misleading indicator of performance for mission-oriented systems. The MTTF is given by

$$\text{MTTF} = \int_0^\infty R\, dt \tag{11.7}$$

That is, MTTF computation considers reliability over the interval $[0, \infty]$ which includes period after the crossover. For a mission-oriented system, the reliability function is of concern over the period $[0, T]$ which may be before the crossover. The crossover point can be determined by equating the reliability of simplex and TMR systems,

$$R_{\text{TMR}} = R$$

that is,

$$R^2(3-2R)=R$$

$$2R^2-3R+1=0$$

or

$$R=1,1/2$$

Now $R=1$ corresponds to $T=0$ and $R=1/2$ corresponds to

$$e^{-\lambda T}=0.5$$

that is,

$$\lambda T=0.693$$

TMR with Repair. In the above analysis of TMR, the three units are assumed essentially nonrepairable. This type of analysis is applicable to situations like space exploration. In the ground-based applications the failed unit can be repaired, however, and the configuration restored to TMR if another unit has not failed meanwhile. The reliability equation for such an application can be derived using the state transition diagram shown in Figure 11.3, where λ and μ are the failure and repair rates of each unit. When a unit fails, it can be repaired and restored. The process of restoration (synchronization, etc.) is assumed to be accomplished with probability 1. State 3, where two units are failed, is made an absorbing state by not allowing repair from this state. Under this condition, the reliability of the TMR (TMRR) is

$$R_{\text{TMRR}}=P_1(T)+P_2(T) \tag{11.8}$$

where $P_i(t)=$ probability of the system being in state i at time t.

The state differential equations for Fig. 11.3 are

$$\dot{P}_1(t)=\mu P_2(t)-3\lambda P_1(t) \tag{11.9}$$

$$\dot{P}_2(t)=3\lambda P_1(t)-(2\lambda+\mu)P_2(t) \tag{11.10}$$

$$\dot{P}_3(t)=2\lambda P_2(t) \tag{11.11}$$

Assuming $P_1(0)=1.0$ and taking Laplace transform of (11.9)–(11.11)

$$sP_1(s)-1=\mu P_2(s)-3\lambda P_1(s) \tag{11.12}$$

$$sP_2(s)=3\lambda P_1(s)-(2\lambda+\mu)P_2(s) \tag{11.13}$$

$$sP_3(s)=2\lambda P_2(s) \tag{11.14}$$

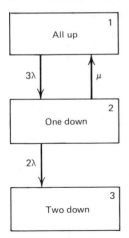

Figure 11.3 State transition diagram of TMR with repair.

From (11.12) and (11.13)

$$P_1(s) = \frac{s + 2\lambda + \mu}{s^2 + (5\lambda + \mu)s + 6\lambda^2} \qquad (11.15)$$

and

$$P_2(s) = \frac{3\lambda}{s^2 + (5\lambda + \mu)s + 6\lambda^2} \qquad (11.16)$$

Taking inverse Laplace of (11.15) and (11.16)

$$P_1(t) = \frac{1}{r_1 - r_2} \left[(2\lambda + \mu)(e^{r_1 t} - e^{r_2 t}) + r_1 e^{r_1 t} - r_2 e^{r_2 t} \right] \qquad (11.17)$$

and

$$P_2(t) = \frac{3\lambda}{r_1 - r_2} \left[e^{r_1 t} - e^{r_2 t} \right] \qquad (11.18)$$

where

$$r_1, r_2 = \frac{-(5\lambda + \mu) \pm \sqrt{\lambda^2 + \mu^2 + 10\lambda\mu}}{2}$$

Substituting into (11.8)

$$R_{\text{TMRR}} = \frac{1}{r_1 - r_2} \left[(5\lambda + \mu)(e^{r_1 T} - e^{r_2 T}) + r_1 e^{r_1 T} - r_2 e^{r_2 T} \right] \qquad (11.19)$$

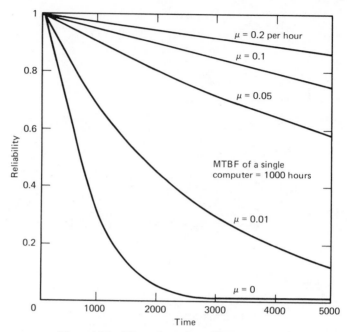

Figure 11.4   Effect of repair on TMR reliability.

Equation 11.19 gives the reliability of a TMR configuration with repair. Figure 11.4 shows the effect of repair rate parameter on the TMR reliability. If there is no repair, then substituting $\mu=0$ into (11.19)

$$R_{\text{TMRR}} = 3e^{-2\lambda T} - 2e^{-3\lambda T}$$

$$= R^2(3-2R)$$

$$= R_{\text{TMR}}$$

TMR With Repair and Common Mode Failures.   Analysis of TMR up to this point is based on the independent failures of the units comprising the TMR. In practice, some faults like those due to external environment and software bugs can cause common mode failures. The state transition diagram for this situation is shown in Figure 11.5. Let

$\lambda=$ failure rate of a single unit $= 1/U_1$
$U_1 =$ mean up time of a single computer
$\lambda_c =$ rate for common mode failures
    $= 1/U_c$
$U_c =$ mean operating time between consecutive common mode failures
$\mu =$ repair rate of a computer
    $= 1/(\text{mean time to repair})$

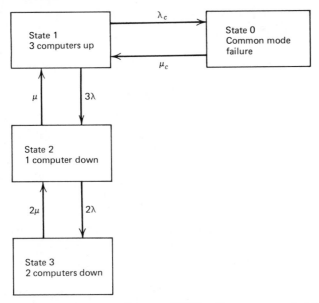

Figure 11.5 State transition diagram of TMR with common mode failure.

μ_c = repair rate for common mode failure
= 1/(mean time to repair common mode failure)

The frequency balance equations [14] for Figure 11.5 are given by (11.20)–(11.23),

$$(3\lambda+\lambda_c)P_1=\mu P_2+\mu_c P_0 \tag{11.20}$$

$$\lambda_c P_1=\mu_c P_0 \tag{11.21}$$

$$(2\lambda+\mu)P_2=3\lambda P_1+2\mu P_3 \tag{11.22}$$

$$2\mu P_3=2\lambda P_2 \tag{11.23}$$

From (11.22) and (11.23)

$$P_1=\frac{\mu}{3\lambda}P_2 \tag{11.24}$$

Mean up time of TMR is given by [14]

$$U_{TMR}=\frac{P_1+P_2}{\lambda_c P_1+2\lambda P_2}$$

$$=\left(\frac{\mu}{3\lambda}+1\right)\bigg/\left(\frac{\mu\lambda_c}{3\lambda}+2\lambda\right)$$

$$=\frac{x_r U_1^2}{U_1+6x_r d}+\frac{3x_r d U_1}{U_1+6x_r d} \tag{11.25}$$

where d = mean repair time = $1/\mu$
$\quad\quad x_r = U_c/U_1$

The mean up time of TMR depends upon x_r. For example, if

$$x_r = 1, \text{ that is, } U_c = U_1$$

and

$$d \ll U_1$$

Then from (11.25)

$$U_{\text{TMR}} \simeq U_1$$

If U_c is comparible with U_1, little is gained by triple redundancy as far as reliability is concerned. If, however, $x_r \gg 1$; in the limiting case of $x_r \to \infty$ meaning no common mode failures

$$U_{\text{TMR}} = \frac{U_1^2}{6d} + \frac{U_1}{2} \tag{11.26}$$

It is worth noting that U_{TMR} is sensitive to the square of U_1. Therefore if U_1 is doubled, U_{TMR} is approximately quadrupled.

11.5.3 NMR Configuration

An NMR (N-tuple Modular Redundant) system consists of $N = 2n + 1$ replicated units feeding a $(n + 1)$-out-of N voter. For a majority among N units, at least $n + 1$ units must survive for successful operation. Neglecting the effect of compensating failures, the reliability of an NMR configuration is given by (11.27).

$$R_{\text{NMR}} = \sum_{i=0}^{n} \binom{N}{i} (1 - R)^i R^{N-i} \tag{11.27}$$

where $\binom{N}{i} = N!/(N-i)!i!$. If R_{NMR} is plotted as a function of normalized time λT, it is seen that as in the case of TMR, R_{NMR} is greater than simplex reliability before the crossover point and less than simplex after the crossover point. This effect is accentuated as N increases and in the limiting case of $N \to \infty$, the R_{NMR} is unity before crossover point and zero afterward.

11.5.4 Standby Redundancy

Several functionally identical units are employed in standby redundancy. Some of the units perform active function whereas the others are in a standby mode, waiting to be switched in if one of the units fails. Bouricious et al. [4] made an important observation that the reliability of stand-by redundant configurations is sensitive to coverage. Coverage can be defined as the probability of successful sensing, switching, and recovery. It can also be defined as the fraction of all possible faults in the computer system from which the system can recover by reconfiguration.

System With One Spare. This system consists of two identical computing modules. One of the modules is active and the other is in a stand-by mode. Let

λ_a = failure rate of the active module
λ_s = failure rate of the standby module
μ = repair rate of either module
c = coverage
$\bar{c} = 1 - c$

The time between failures and the repair time are assumed exponentially distributed. The state transition diagram of this system is shown in Figure 11.6 where A and S stand for active and stand-by, and G and F stand for good and failed. Both the modules are good in state 1. When the standby fails, the system enters state 3. When the active unit fails, there are two possible resulting states. If the failure of the active unit is sensed, the standby module is switched in and there is a successful recovery, the system enters state 2 where the standby now becomes active. In the case of unsuccessful recovery, both the units go down and there is system failure—state 4. The transition rate matrix for Figure 11.6 is

$$A \equiv \begin{bmatrix} -(\lambda_a+\lambda_s) & c\lambda_a & \lambda_s & \bar{c}\lambda_a \\ \mu & -(\mu+\lambda_a) & 0 & \lambda_a \\ \mu & 0 & -(\mu+\lambda_a) & \lambda_a \\ 0 & \mu & \mu & -2\mu \end{bmatrix}$$

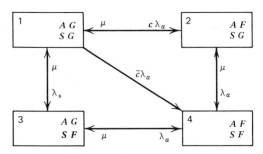

Figure 11.6 State transition diagram of a standby system.

Assuming the process starts in state 1, the mean time to first failure (MTTFF) can be obtained [14] from (11.28)

$$\text{MTTFF} = (1 \ 0 \ 0)[-Q]^{-1}U \tag{11.28}$$

where $Q =$ is the matrix obtained from A by deleting the row and column corresponding to state 4 indicating system failure
$U =$ is a unit column vector

Now

$$-Q = \begin{bmatrix} \lambda_a + \lambda_s & -c\lambda_a & -\lambda_s \\ -\mu & \mu + \lambda_a & 0 \\ -\mu & 0 & \mu + \lambda_a \end{bmatrix}$$

Denoting the element (ij) of the inverse of $-Q$ by m_{ij},

$$m_{11} = \frac{(\mu + \lambda_a)^2}{\Delta}$$

$$m_{12} = c\lambda_a \frac{(\mu + \lambda_a)}{\Delta}$$

$$m_{13} = \lambda_s \frac{(\mu + \lambda_a)}{\Delta}$$

where $\Delta = |-Q|$

$$= \lambda_a(\mu + \lambda_a)(\bar{c}\mu + \lambda_a + \lambda_s)$$

Using (11.28)

$$\text{MTTFF} = m_{11} + m_{12} + m_{13}$$

$$= \frac{\mu + \lambda_s + \lambda_a(1 + c)}{\lambda_a(\bar{c}\mu + \lambda_a + \lambda_s)} \tag{11.29}$$

If $\lambda_a = \lambda_s$, (11.29) reduces to (11.30)

$$\text{MTTFF} = \frac{\mu + \lambda_a(2 + c)}{\lambda_a(\bar{c}\mu + 2\lambda_a)} \tag{11.30}$$

If, however, $\lambda_s \ll \lambda_a$, then

$$\text{MTTFF} = \frac{\mu + \lambda_a(1 + c)}{\lambda_a(\bar{c}\mu + \lambda_a)} \tag{11.31}$$

Table 11.2   *MTTFF as a function of c*

$\lambda_a = 0.00025/\text{hour}$
$\mu\ = 0.25/\text{hour}$

	MTTFF (hours)	
c	For $\lambda_s = \lambda_a$	For $\lambda_s = 0$
1	2,006,000	4,008,000
0.99	334,330	364,360
0.98	182,360	190,853
0.95	77,150	78,584

As an example assume

$$\lambda_a = 0.00025/\text{hour}$$

and

$$\mu = 0.25/\text{hour}$$

The MTTFF for this system is shown as a function of c in Table 11.2. It can be seen that the MTTFF is very sensitive even to small variation in c. The analysis of this system with constant repair time is reported in reference 6

11.6   MODELING TRANSIENT FAULTS

A transient fault is of limited duration and disappears some time after its arrival [10]. During this brief period of its stay, it can, however, alter the contents of registers and interfere with the normal sequence of program execution. The transient fault, although of limited duration, can cause incorrect execution of the program and therefore result in computer system failure. Redundancy can be used to recover from the effect of transient faults. Redundancy is used to detect as well as provide recovery. The detection of fault can be made by comparing the outputs of two or more modules. The fault-free computer is also used to assist in restoring altered data and program and also synchronizing the recovering computer.

Reference [10] describes the modeling of the transient faults in a TMR configuration. The transient faults are assumed to arrive with an average rate λ_t, assumed constant over the system life. The duration of the transient faults is assumed to be distributed exponentially, given by

$$f_D = re^{-rt} \tag{11.32}$$

The recovery procedure consists of three steps:

1. *Detection.* Faults are detected by comparison, the time between comparisons being T_c. The time between the occurrence and detection of a fault is a random variable with an assumed probability density function $\delta \exp(-\delta t)$.
2. *Delay.* A certain time T_D is allowed between the detection of fault and initiation of recovery. This time is designed for the transient to die out.
3. *Recovery.* A certain time T_R is allowed to accomplish the recovery procedure.

If all the computers are fault-free, and now a transient occurs in one of the computers, then a transient recovery procedure will be initiated with the following possible outcomes:

1. Recovery is successful and therefore the computers are again fault-free.
2. The transient may be long enough to continue into the recovery period and be mistaken for a permanent fault. In this case this computer will be assumed faulted.
3. A previously fault-free computer may experience a fault during the recovery period resulting in TMR system failure.

The system is assumed to fail when a computer experiences a permanent or transient fault when another computer is having a permanent fault or is recovering from a transient. The probability of a system failing before time T is derived in reference 10 and is,

$$= a \left\{ \frac{1 - e^{-3bT}}{b} - \frac{3e^{-2cT}}{3b - 2c} \left[1 - e^{-(3b - 2c)T} \right] \right\}$$

$$+ \frac{\bar{L}\lambda_t}{b} \left[\frac{2c + \delta(1 - e^{-2cT_r})}{\delta + 2c} \right] (1 - e^{-3bT}) \qquad (11.33)$$

where $a = \lambda + L\lambda_t$
        $b = \lambda + (1 - c_T)\lambda_t$
        $c = \lambda + \lambda_t$
        $c_T = $ transient coverage
            $= $ Probability of recovery from a transient, given a transient occurs.
        $L = $ Probability of transient fault being interpreted as a permanent fault.

$$= 1 - \delta e^{-cT_r} \left[\frac{1}{\delta + c} - \frac{e^{-rT_D}}{c + r + \delta} \right]$$

$$\overline{L} = 1 - L$$
$$T_r = T_D + T_R$$

REFERENCES

1. Avizienis, A., G. C. Gilley, F. P. Mathur, D. A. Rennels, J. A. Rohr, and D. K. Rubin, "The STAR (Self-Testing-And-Repairing) Computer: An Investigation of the Theory and Practice of Fault-Tolerant Computer Design," *IEEE Trans. Comput.*, **C-20**, 1312–1321 (1971).

2. Avizienis, A., "Modeling and Evaluation of the Reliability of Computer Systems," *Third National Reliability Symposium*, Perros-Guirec, France Sept. 14–17, 1976.

3. Avizienis, A., "Fault-Tolerant Computing: Progress, Problems, AND Prospectus," Inf. Process '77, Proceedings of IFIP Congress, Toronto, Aug. 8–12, 1977. Published by North-Holland Publ. Co. (IFIP Congress series, vol. 7), Amsterdam and New York, NY, 1977, pp. 405–420.

4. Bouricius, W. G., W. C. Carter, and P. R. Schneider, "Reliability Modeling Techniques For Self-Repairing Computer Systems," *Proceedings of ACM National Conference*, ACM, New York, 1969.

5. Carter, W. C. and W. G. Bouricius, "A Survey of Fault Tolerant Computer Architecture and Its Evaluation," *Computer*, **4**, 9–16 (1971).

6. Chow, D. K., "Reliability of Some Redundant Systems with Repair," *IEEE Trans. Reliab.*, **R-22**, 223–228 (1973).

7. Goldberg, J., "A Survey of the Design and Analysis of Fault-Tolerant Computers," *Reliability and Fault Tree Analysis*, Theoretical and Applied Aspects of System Reliability and Safety Assessment, SIAM, Philadelphia, 1975, pp. 687–731.

8. Mathur, F. P. and A. Avizienis, "Reliability Analysis and Architecture of a Hybrid-Redundant Digital System: Generalized Triple Modular Redundancy With Self-Repair," 1970 Spring Joint Computer Conference, *AFIPS Conference Proceedings*, Vol. 36, Montvale, NJ, May 1970, pp. 375–387.

9. Mathur, F. P. and P. T. de Sousa, "Reliability Models of NMR Systems," *IEEE Trans. Reliability*, **R-24**, 108–113 (1975).

10. Merryman, P. M. and A. Avizienis, "Modeling Transient Faults in TMR Computer Systems," *Proceedings 1975 Annual Reliability and Maintainability Symposium*, IEEE, New York, 1975.

11. Von Neumann, J., "Probabilities Logics and the Synthesis of Reliable Organisms From Unreliable Components," In: *Automata Studies*, Princeton University Press, Princeton, NJ, 1956.

12. Nerber, P. O., "Power-Off Time Impact on Reliability Estimates," *IEEE Int. Conv. Rec.*, Part 10, March 1965, pp. 1–8.

13. Short, R. A., "The Attainment of Reliable Digital Systems Through the Use of Redundancy—A Survey," *Computer Group News*, **2**, 2–17 (March 1968).

14. Singh, C. and R. Billinton, *System Reliability Modelling and Evaluation*, Hutchinson, London, 1977.

15. Yourdon, E., "Reliability Measurements For Third Generation Computer Systems," *Proceedings 1972 Annual Reliability and Maintainability Symposium*, IEEE, New York, 1972.

16. Yourdon, E., "Reliability of Real-Time Systems—Part 1—Different Concepts of Reliability," *Modern Data*, **5**, 36–42 (Jan. 1972).

17. Yourdon, E., "Reliability of Real-Time Systems—Part 2—The Causes of System Failures," *Modern Data*, **5**, 50–56 (Feb. 1972).

18. Yourdon, E., "Reliability of Real-Time Systems—Part 3—The Causes of System Failures Continued," *Modern Data*, **5**, 36–40 (March 1972).

19. Yourdon, E., "Reliability of Real-Time Systems—Part 4—Examples of Real-Time System Failures," *Modern Data*, **5**, 52–57 (April 1972).

20. Yourdon, E., "Reliability of Real-Time Systems—Part 5—Approaches to Error Recovery," *Modern Data*, **5**, 38–52 (May 1972).

Appendix

Three-State Device Networks

All derivations in the appendix are based on the binomial expansion $(P+q)^n$. In the case of three-state device structures, the expansion is modified to $(P+q_o+q_s)^n$, where P is the component probability of success, q_o open failure probability, q_s short failure probability, and n is the number of independent elements.

A.1 SERIES STRUCTURE

A.1.1 Reliability Expression

Let $n=2$. Therefore

$$(P+q_o+q_s)^2=1$$

Thus

$$P^2+2Pq_o+2Pq_s+q_o^2+2q_oq_s+q_s^2=1$$

The number of state combinations for $(n=2)=3^2=9$. The state combination truth table may be represented as follows:

Table A1.

NN	NO	NS
ON	OO	OS
SN	SO	SS

where $N=$ the normal mode of the device (success)
$O=$ the open mode failure state
$S=$ the short mode failure state

The reliability terms, by inspection of the truth table, Table A1, are

$$PP+Pq_s=Pq_s=P^2+2Pq_s$$

Since
$$P = 1 - q_s - q_o, \tag{A.1}$$

$$R = \left(1 - q_o^2\right) - q_s^2 \tag{A.2}$$

Now let $n = 3$:
Therefore

$$(P + q_o + q_s)^3 = 1$$

Thus

$$P^3 + 3P^2 q_o + 3P^2 q_s + 3Pq_0^2 + 3Pq_s^2 + 6Pq_o q_s + q_o^3 + 3q_o^2 q_s + 3q_s^2 q_o + q_s^3 = 1$$

Number of state combinations for $(n = 3) = 3^3 = 27$. The new state combinations truth table is as shown in Table A2.

Table A2.

NNN	SSS	OOO
NOO	SOO	OSS
OON	OOS	SSO
NON	OSO	ONN
ONO	SOS	NNO
NSS	SNN	NSO
SSN	NNS	NOS
SNS	NSN	SNO
SON	OSN	ONS

The reliability terms by inspection of Table A2

$$PPP + Pq_s q_s + Pq_s q_s + Pq_s q_s + PPq_s + PPq_s + PPq_s$$

therefore

$$R = P^3 + 3Pq_s^2 + 3P^2 q_s \tag{A.3}$$

By using (A.1) and (A.3):

$$R = (1 - q_o)^3 - q_s^3 \tag{A.4}$$

Obviously, from (A.2) and (A.4), the general equation for the reliability of the series structure can be written as follows. For identical components:

$$R = (1 - q_o)^n - q_s^n \tag{A.5}$$

and in the case of nonidentical components:

$$R= \prod_{i=1}^{n} (1-q_{oi}) - \prod_{i=1}^{n} q_{si} \qquad (A.6)$$

A.1.2 Probabilities of Failure

At $n=2$. From Table A1 open mode failure terms

$$q_o P + P q_o + q_o q_o + q_s q_o + q_o q_s$$

Since

$$P = 1 - q_o - q_s \qquad (A.7)$$

Therefore the series system open mode failure probability

$$Q_o = 1 - (1 - q_o)^2 \qquad (A.8)$$

and similarly for short-mode failure probability

$$Q_s = q_s^2 \qquad (A.9)$$

At $n=3$. According to Table A2, open mode failure terms

$$3 P q_o^2 + 3 P^2 q_o + 6 P q_o q_s + 3 q_o^2 q_s + 3 q_s^2 q_o + q_o^3 \qquad (A.10)$$

By substituting for P in (A.10), open-mode failure probability

$$Q_o = 1 - (1 - q_o)^3 \qquad (A.11)$$

and in the same way for short-mode failure

$$Q_s = q_s^3 \qquad (A.12)$$

In the case of identical components, according to (A.8) and (A.11), the general form of open mode system failure

$$Q_o = 1 - (1 - q_o)^n \qquad (A.13)$$

Similarly, for nonidentical components

$$Q_o = 1 - \prod_{i=1}^{n} (1 - q_{oi}) \qquad (A.14)$$

In the case of short mode system failure for identical components

$$Q_s = q_s^n \tag{A.15}$$

and for the nonidentical elements case

$$Q_s = \prod_{i=1}^{n} q_{si} \tag{A.16}$$

A.2   PARALLEL STRUCTURE

A.2.1   *Reliability Expression*

Let the number of parallel elements $m = 2$

$$\therefore (P + q_o + q_s)^2 = 1$$

Thus

$$P^2 + 2Pq_o + 2Pq_s + q_o^2 + 2q_oq_s + q_s^2 = 1$$

The reliability terms with inspection From Table A1

$$PP + q_oP + Pq_o = p^2 + 2Pq_o \tag{A.17}$$

Since $P = 1 - q_s - q_o$, therefore

$$R = (1 - q_o)^2 - q_s^2 \tag{A.18}$$

At $m = 3$

$$\therefore (P + q_o + q_s)^3 = 1$$

Thus

$$P^3 + 3P^2q_o + 3P^2q_s + 3Pq_o^2 + 3Pq_s^2 + 6Pq_oq_s + q_o^3 + 3q_o^2q_s + 3q_oq_s^2 + q_s^3 = 1$$

By inspecting Table A2, system reliability

$$= P^3 + 3q_o^2P + 3P^2q_o \tag{A.19}$$

Replace P with (A.7); therefore

$$R = (1 - q_s)^3 - q_o^3 \tag{A.20}$$

With the aid of (A.18) and (A.20), the general system reliability formula for identical elements connected in the parallel configuration becomes as follows:

$$R = (1 - q_s)^m - q_o^m \tag{A.21}$$

The above equations for nonidentical elements may be rewritten as

$$R = \prod_{i=1}^{m} (1 - q_{si}) - \prod_{i=1}^{m} q_{oi} \tag{A.22}$$

A.2.2 Failure Probabilities

For $m = 2$. Collected short-mode failure terms from Table A1

$$= q_s P + q_s q_o + q_s q_s + q_o q_s + P q_s = q_s^2 + 2 P q_s + 2 q_o q_s$$

Thus:

$$\therefore Q_s = 1 - (1 - q_s)^2 \tag{A.23}$$

Similarly, for open-mode failure

$$Q_o = q_o^2 \tag{A.24}$$

At $m = 3$. Short-mode system failure terms from Table A2

$$= q_s^3 + 3 P q_s^2 + 6 P q_o q_s + 3 q_s q_o^2 + 2 q_o q_o^2 + 3 q_s q_o^2 + 3 q_s^2 q_o$$

Thus

$$Q_s = 1 - (1 - q_s)^3 \tag{A.25}$$

Similarly, in the case of open mode failure

$$Q_o = q_o^3 \tag{A.26}$$

As seen from (A.23) and (A.25), the short-mode failure equation for identical elements can be generalized as follows:

$$Q_s = 1 - (1 - q_s)^m \tag{A.27}$$

Similarly, for the nonidentical elements case

$$Q_s = 1 - \prod_{i=1}^{m} (1 - q_{si}) \tag{A.28}$$

The corresponding generalized open mode system failure probability

$$Q_o = q_o^m \qquad \text{identical components} \qquad (A.29)$$

and

$$Q_o = \prod_{i=1}^{m} q_{oi} \qquad \text{nonidentical components} \qquad (A.30)$$

Index